God Made THE WORLD

Science/Worldview | 5-7 Grade

Author: Kevin Swanson
BS, Mechanical Engineering, California Polytechnic State University,
MDiv, Southern California Center for Christian Studies

Consultant: Mark Abbotoy
BS, Mechanical Engineering, Clarkson University,
MS, Applied Statistics and Research Methods, University of Northern Colorado

Generations
PASSING ON THE FAITH

Copyright © 2020 by Generations. All rights reserved. No part of this book may be reproduced in any form or by any means without permission in writing from the publisher.

Printed in the United States of America.

2nd Printing, 2021.

ISBN: 978-1-7350719-4-7

Cover Design: Justin Turley
Interior Design: Sarah Bryant

Published by:
Generations
19039 Plaza Drive Ste 210
Parker, Colorado 80134
Generations.org

Unless otherwise noted, Scripture taken from the New King James Version®. Copyright © 1982 by Thomas Nelson. Used by permission. All rights reserved.

For more information on this and other titles from Generations, visit Generations.org or call (888) 389-9080.

TABLE OF CONTENTS

Introduction
For the Parent and Teacher 7

Chapter 1
God's Works are Awe-Inspiring! 11

Chapter 2
God Made Glorious Sights 25

Chapter 3
God Made the Earth 63

Chapter 4
God Made Matter 101

GOD MADE THE WORLD

Chapter 5
God Gives Us Energy 141

Chapter 6
God's Design of Motion 177

Chapter 7
God Made Waves—Sound and Light 201

Chapter 8
God Made Dirt and Water 229

Chapter 9
God Gives Us Atmosphere and Weather 259

Endnotes 293

List of Images 297

Index 303

Haifoss Waterfall, Iceland

Great Ocean Road, Victoria, Australia

Introduction
FOR THE PARENT AND TEACHER

This course is intended to produce a paradigm shift in the way the present generation of Christian children understand science. Differing worldviews will inevitably yield radically different approaches to science. Methods of study, purposes of study, and the content of study will vary greatly depending on one's presuppositions and beliefs. A biblical worldview perspective of science will always place God at the center. God is personal, and He is the source of all creation. His fingerprints are seen everywhere throughout this wide world. All knowledge of His creation will inevitably, constantly reveal His power and glory unless great pains are taken to willfully and continually suppress these.

God's attributes shine brilliantly in His created world. Thus, the reader will find this text always delighting in the revelation of God's genius, wisdom, power, and goodness everywhere manifest in creation. Science provides us a perfect opportunity to maximize on the great purpose for all of life. Indeed, the purpose of science is to enjoy God in the context of His creation and to praise Him for His marvelous works!

The study of science is deeply personal because it is revelatory of the personality of God. The pieces of His creation are not random accidents in a chance universe! They are the careful design of a personal, loving, wise, and all-powerful Creator! This knowledge should inspire the student to become fully engaged while rejoicing, delighting, praising God, and thanking Him for the awesomeness and manifold blessings of His creation!

Throughout this course, we glory in the incomparable wisdom of God! With effervescent delight, we inform the student of the bounds of human knowledge in respect to the atom, nature's forces, the nature of light, black holes, and a hundred other mysteries of the universe. We are not afraid to say that we can't explain this or that. We cheerfully inform the student that the greatest scientists in the world cannot begin to comprehend the deep mysteries of God's wisdom in His creation. We fall on our faces in awe-filled worship! This is the only way to avoid the academic pride, scientific hubris, and lack of the fear of God that has ruined science and education in our humanist age.

This course also provides extensive devotional reading from Scripture throughout its text, for all Christian education must retain the Word of God as "a frontlet" before our eyes and our children's eyes (Deut. 6:7-9). We offer many opportunities for prayer and the singing of praise. These elements are core to a Christian view of science. As the teacher/parent disciples the child in the study of God's creation, we hope and pray that the student will form a Godward view of science and all of life.

If education is to be truly effective, the student must be constantly informed of the vital importance and objective of the study. Hardly a page of text should go by without the student realizing the high significance and purpose of what he is studying. For the Christian, the purpose of science is absolutely clear. Studying God's works in creation should lead us to praise and worship Him and to take proper dominion as good stewards over God's world. We glory in His nature and His works, and we seek to fulfill our role in ruling over the natural world (as the Lord commanded man to do at the beginning).

INTRODUCTION: FOR THE PARENT AND TEACHER

Perpetually spewing out disconnected and purposeless facts to be gathered into a child's brain does little for retention. However, if a clear integration of praise and life application is found on every page, the material will be much better retained. When science instruction is given meaning and purpose on every page of text and in every minute of the class, the student will be much more likely to retain the material and apply it in a meaningful way in his life.

The "Do" sections contained at the end of each chapter are not intended to serve as typical "laboratory" experiences or hypothetical exercises. Rather, they are intended to serve as real-life application of the science conveyed in each chapter. We want the students to actually practice the science they learn for the real benefit of their family, their community, and their personal economy. Although many of the practical projects suggested are simple and easy, we would recommend taking on only one *major* project for the academic year. Teacher and parent involvement is highly recommended for these projects.

For 5th to 7th grade students, Generations here introduces a biblical worldview into basic astronomy, chemistry, and physics in the most winsome way possible. Captured in this introduction to science are the most amazing facts and the most interesting facets of God's creation. Efforts are made to explain difficult concepts at a 10-12-year-old level without losing the substance of the scientific meaning. Our goal is to create a work in which not a single paragraph is uninteresting, vague, or overly complex.

This text introduces naturally occurring, providential events such as earthquakes, volcanoes, tornadoes, and hurricanes. Yet the goal is always to point the student back to our sovereign God in right reverence and worship. Then we offer helpful, relevant, and interesting life (and spiritual) applications for each disastrous scenario—preparing each young student to respond rightly in faith, wisdom, and faithful stewardship to these awe-inspiring acts of God.

Tian Shan Mountain Range, Asia

Chapter 1
GOD'S WORKS ARE AWE-INSPIRING!

> The works of the LORD are great, studied by all who have pleasure in them. (Psalm 111:2)

An ant crawls across a great work of art. The beautiful painting hangs in an art museum in Paris, France. The people who have come to the museum to see the art are amazed at the works of art. But the ant doesn't consider the beauty of the art, nor does he think about the artist who painted the piece. He does not appreciate the great work and he is definitely not amazed by it.

When you look around at God's world, do you act like the ant in the art museum? Do you observe God's amazing creation, and fail to enjoy it? Do you stand back in awe and amazement at the great work of art that God has produced?

God made this world, and He takes care of His world. What you see all around you are the works of God. He made all of this. The works of God are not very great for those who will not study them. Some people are bored when they study the creation because they are not delighting in them. First, we must study God's world. Because we love God and we can see His handiwork all around us, we will delight in all of it.

Now, we shall study God's world and delight in it throughout this book. We do love God, and we want to learn more about Him and His wonderful works. We want to see His hand in this world.

GOD MADE THE WORLD

Logs in a Pine Forest

The study of God's world takes time. If we are interested in all that He has done, we will take the time to study them. These investigations also call for focus. We cannot be easily distracted with other things. You will not see the beauty of God's creation, or the complex nature of it, or the many features of it, unless you study it for a long time.

Some children study their science lessons just so they will pass the test. Or, they study science because their mothers told them to. They are not very interested in God's amazing world. But that is not the way that we will study the creation. We will study it, until we see the wonder of it. Then, we will praise God for it.

This is My Father's World

The earth is the LORD's, and all its fullness, the world and those who dwell therein. For He has founded it upon the seas, and established it upon the waters. (Psalm 24:1-2)

God made the world. He made everything on this earth, and everything in the sky. He lightened up our whole world with the sun and the moon. He created oceans, lakes, rivers, mountains, water, air, oil, sandstone, and dirt—plenty of materials to take care of us on this earth. He gave us plenty of wood so we could build things for our own use. Man

CHAPTER 1: GOD'S WORKS ARE AWE-INSPIRING!

builds houses. He makes tables and chairs, but God makes the wood from which man builds things. The wood is called **raw material**. God made all the raw material in the universe. This is called "God's work of creation."

God made all the materials you see around you. Take a look at yourself. Look at your hands and feet. God made your hands so you can pick up things quite easily. He made your eyes so that you can see the things that you are picking up. He made your mind so you can think about what you are going to pick up next. The most complicated and the most wonderful creation in this world is man—you, your family, and your friends.

God makes life—plant, animal, and human life. He brings new life into the world when babies are born.

The Days of Creation

In the beginning God created the heavens and the earth. (Genesis 1:1)

The first chapter in Genesis tells us how God created the world and

View of Earth from the Moon

GOD MADE THE WORLD

everything within it in six days.

Creation Day 1 (Genesis 1:1-5)

God created the heavens and the earth. The heavens mean space outside of the material earth. The earth is the globe with some sort of matter.

Creation Day 2 (Genesis 1:6-8)

God created the sky with clouds and moisture in the air.

Creation Day 3 (Genesis 1:9-13)

God created dry land and the oceans. God made all plant life.

Creation Day 4 (Genesis 1:14-19)

God created the moon, the sun, and the stars in the sky.

Creation Day 5 (Genesis 1:20-23)

God made fish, birds, and flying insects. He gave these animals the ability to reproduce (or have babies).

Creation Day 6 (Genesis 1:24-31)

God created all the animals that live on dry land. And finally, God made man in His own image.

Andromeda Galaxy

CHAPTER 1: GOD'S WORKS ARE AWE-INSPIRING!

In this study, we will look at the non-living things which God made during the first four days of creation.

God's Creation is More Amazing Every Year We Study It

Praise the LORD!
Praise the LORD from the heavens;
Praise Him in the heights!
Praise Him, all His angels;
Praise Him, all His hosts!
Praise Him, sun and moon;
Praise Him, all you stars of light!
Praise Him, you heavens of heavens,
And you waters above the heavens!
Let them praise the name of the LORD,
For He commanded and they were created.
He also established them forever and ever;
He made a decree which shall not pass away.
(Psalm 148:1-6)

Milky Way Galaxy

What can we say about all of these things God has made? We find that so much of this world is beautiful. We find that His creation is useful. As we study God's creation, we learn that it is very complicated. Some **scientists** have studied the human body for sixty years, and they still do not understand it very well. Over thousands of years, men have studied the human body, but there is still so much more to learn. This means that God is very smart. He is all wise.

It took a long time for man to learn about the universe. A Greek scientist named Democritus suggested that the stars in the night sky might be thousands of suns making up a galaxy. This we know as the Milky Way. Two other galaxies separate from the Milky Way were discovered around 1000 AD, fully 5,000 years after God created the world. By 1920, **astronomers** thought the universe was only 30,000 light years across. Still at this time these modern scientists thought there was only one galaxy in the sky—our

GOD MADE THE WORLD

Milky Way.[1] After the development of radio telescopes and other instruments, scientists have learned much more about the universe. Now, they say that the universe is millions of times larger than they thought it was 100 years ago. You can only see about 2,500 stars from our galaxy by your own eyes. Scientists tell us they have found between 100 and 300 billion stars in the Milky Way. By 1960, astronomers had found 30,000 galaxies. But now they tell us there are between 100 billion and 1 trillion galaxies in the sky.

We have learned that the universe is far bigger, far more awe-inspiring than we thought it was 100 years ago. Sadly, many astronomers do not give God the glory anymore. They have learned so much more about God's awesome universe, but they do not praise Him. In a survey done with these astronomers, 71% of them say that they do not believe in God. Among the general public in America, only 17% would say that they do not believe in God.[2]

Everything belongs to God. He lends His materials to us for a while. We want to use them for good purposes. He wants us to take good care of His world. He wants us to make helpful things out of the things which He has made. Sometimes, people who own businesses will hire men to take care of the business for them. These are called **stewards**. God has appointed us as stewards to take care of His world.

God's Works of Providence

For by [the LORD Jesus] all things were created that are in heaven and that are on earth, visible and invisible, whether thrones or dominions or principalities or powers. All things were created through Him and for Him. And He is before all things, and in Him all things consist. (Colossians 1:16-17)

CHAPTER 1: GOD'S WORKS ARE AWE-INSPIRING!

About 6,000 years ago, God created the world in six days. He made the earth, the oceans, the sun, moon, and stars, the plants, animals, and humans. He made it all out of nothing. We make chairs and tables out of wood, but God made the whole world out of nothing. This shows the amazing power and wisdom of God.

Now, God did not create the world and leave it alone. He is still very involved with this world. He takes care of this world by what we call "God's Works of **Providence**." God brings about changes in His creation. About 4,500 years ago, He brought a worldwide flood that changed the whole surface of the world. The mountains and the valleys, the rivers and the plains were formed by the worldwide flood.

When you blow on a lighted match, you cause a little wind to blow out the flame. God causes the winds to blow all around the earth. He brings big storms and earthquakes. When you throw dice on a game board, God causes the dice to turn over a certain number of times. He determines the roll of the dice, whether the outcome is a two or a six. God is in control of everything that happens in the world. When a sparrow falls to the ground, God has made sure that it happens. He takes care of every part of His creation. Everything turns out exactly the way He wants it to be.

Noah's Ark and the Worldwide Flood

The lot is cast into the lap, but its every decision is from the LORD. (Proverbs 16:33)

Are not two sparrows sold for a copper coin? And not one of them falls to the ground apart from your Father's will. But the very hairs of your head are all numbered. Do not fear therefore; you are of more value than many sparrows. (Matthew 10:29-31)

Most importantly, God takes care of us. He makes sure that enough rain falls

GOD MADE THE WORLD

on the fields each year to provide food for 7 billion people in the world. He is especially careful to watch out for His own children—those who do His will and love His Son.

Why God Made the World

For of Him and through Him and to Him are all things, to whom be glory forever. Amen. (Romans 11:36)

The twenty-four elders fall down before Him who sits on the throne and worship Him who lives forever and ever, and cast their crowns before the throne, saying:
"You are worthy, O LORD,
To receive glory and honor and power;
For You created all things,
And by Your will they exist and were created."
(Revelation 4:10-11)

Why did God make you and all things? God made the world for His own glory. Everything is for His glory. Everything that happens in the world will display God's glory.

Psalm 19 says that the "heavens declare the glory of God and the firmament shows His handiwork." All of

Monarch Butterfly

CHAPTER 1: GOD'S WORKS ARE AWE-INSPIRING!

God's creation speaks to us. It is teaching us about God. All of God's creation shouts out to us, "God is glorious! God is glorious! God is glorious!"

God wants us to see His glorious works, and He wants us to praise Him. This is why we study God's world. There are many people who do not see God's glory. They live in America, in Africa, and in Asia. But God wants all peoples to see His glory. "For the earth will be filled with the knowledge of the glory of the LORD, as the waters cover the sea" (Habakkuk 2:14).

Why do you study God's world? Of course, it is to give God the glory for it. We see the great power of God, who made the strong, warm sun. We see the wisdom of God who created the beautiful butterfly to fly so gracefully through the air. We see the goodness of God in all the plants that yield food for us to eat. Then, we thank Him for all of that. We sing songs of praise to Him because He deserves our praise.

God Made the World for Us

And God said, "See, I have given you every herb that yields seed which is on the face of all the earth, and every tree whose fruit yields seed; to you it shall be for food." (Genesis 1:29)

[The LORD] causes the grass to grow for the cattle,
And vegetation for the service of man,
That he may bring forth food from the earth,
And wine that makes glad the heart of man,
Oil to make his face shine,
And bread which strengthens man's heart.
(Psalm 104:14-15)

GOD MADE THE WORLD

God also made the world for man's use. He gave us plants and trees for our enjoyment and for food. He wants us to build things out of the trees, dirt, and rocks. He wants us to make medicine out of the plants. He wants us to make grape juice out of grapes, and orange juice out of oranges. He wants us to make glass out of sand. Then, He wants us to give Him thanks for all of these good things.

You will soon discover that the creation is very complicated, especially as you study the tiniest parts that make up the world. We can learn more and more about all the things our Creator has made. This is what God wants us to do. We cannot take just one look at God's creation and then figure it all out. We have to study it carefully. We must apply our minds to this task. Some call this **science**. We describe God's world. We identify certain differences between the various things God makes, and we give them all names. We notice that the wind causes the trees, the branches, the leaves, and the grass of the field to sway and jiggle. By watching these things, we learn about the cause (the wind) and the effect (the movement of the leaves). God wants us to study His world. It is a very honorable thing to do so.

"It is the glory of God to conceal a matter, but the honor of kings to find it out." (Proverbs 25:2)

How to Be a Good Scientist

The words of Agur the son of Jakeh...
Surely I am more stupid than any man,
And do not have the understanding of a man.
I neither learned wisdom
Nor have knowledge of the Holy One.
Who has ascended into heaven, or descended?
Who has gathered the wind in His fists?
Who has bound the waters in a garment?
Who has established all the ends of the earth?
What is His name, and what is His Son's name,
If you know? (Proverbs 30:1,2-4)

Old Agur, the man who recorded these words in Proverbs 30, was a good scientist. He records things he has learned about the world and wisdom. From this man, we learn the most important mark of a good scientist.

1. **Be humble.** First of all, as we study God's world, we must be very humble.

CHAPTER 1: GOD'S WORKS ARE AWE-INSPIRING!

Sometimes, smart people will become proud of what they know. But this must never be the case for us. In comparison to God, we are very stupid. We do not know very much. We must not brag about what we know. God never makes mistakes, but we often do make mistakes. We are often wrong in our thinking. We only know a small part of what there is to know.

2. **Open your eyes and look at something for a long time.** Look at different parts of what you are looking at. Try to get a closer look at it. Use a magnifying glass or a microscope. Use a telescope to look at things very far away. Take things apart if you can, and look at each piece.

3. **Write down or draw out what you see.** Describe it. What is its color? What is its shape? Use names to identify the different things that you see. What is its function? How does it work?

4. **Make guesses and draw conclusions.** Remember, you cannot be sure of your

first guesses. Do not become proud of what you have learned.

5. **Test your guesses.** Be willing to change your mind when you discover that you were wrong about some of the guesses you made.

The best scientist is the one who is humble and fears the Lord. He is curious. He explores the world with an excited anticipation. He is never tired of seeing the beauty and the amazing design in God's world. He is always praising God, the great Creator of all these things.

I will bless the LORD at all times;
His praise shall continually be in my mouth.
My soul shall make its boast in the LORD;
The humble shall hear of it and be glad.
Oh, magnify the LORD with me,
And let us exalt His name together.
(Psalm 34:1-3)

Pray
- Thank God for all that He has created. Look around you and find five things that He has created and praise Him for them.
- Pray that God would open your eyes to see the wonders of His creation.
- Pray that you will find great delight in this study, that you would look forward to exploring God's world.

Sing
Doxology

Praise God, from whom all blessings flow;
Praise Him, all creatures here below;
Praise Him above, ye heav'nly host;
Praise Father, Son, and Holy Ghost!

If you do not know the hymn, you may listen to a version of the hymn on the Internet, with supervision, and sing along with it.

Watch
To watch the recommended videos for this chapter, go to **generations.org/GodMadeTheWorld** and scroll down until you find the video links for Chapter 1. Our editors have been careful to avoid films with references to evolution. However, we would still encourage parents or teachers to provide oversight for all internet usage. These videos may not give God the glory for His amazing creative work, so the student and parent/teacher should respond to these insights with prayer and praise.

Saturn

Chapter 2
GOD MADE GLORIOUS SIGHTS

> The heavens declare the glory of God; and the firmament shows His handiwork. Day unto day utters speech, and night unto night reveals knowledge. There is no speech nor language where their voice is not heard. Their line has gone out through all the earth, and their words to the end of the world. (Psalm 19:1-3)

On the fourth day of creation, God made the sun, the moon, and the stars. The sun sits right in the middle of what we call our "solar system," and the earth revolves around the sun. He probably made the other planets which revolve around the sun on the same day. These planets are gigantic bodies of mass, made up of gases, rock, silicon, iron and other metals. Men have taken rocket ships to our moon since 1959, and man's first visit to the moon happened on July 20, 1969. This manned

Heavenly Bodies	Spacecraft Landings
The Earth's Moon	47 times since 1959
Venus	15 times since 1966
Mars	14 times since 1971
Mercury	One crash landing in 2015
Saturn's Moon Titan	One landing in 2005

GOD MADE THE WORLD

flight to the moon was called Apollo 11, with the crew of Americans named Neil Armstrong, Michael Collins, and Edwin E. Aldrin, Jr. As of 2020, man has been able to land on the planets and moons listed in the chart on the previous page.

For 6,000 years, man could see some of the planets close to the earth as just tiny lights glimmering in the sky. It has only been in the last sixty years that man has verified the existence of these heavenly bodies by visiting them and sending back video recordings. Our earth is huge. But now, scientists have verified that there are even larger planets than ours, hundreds of millions of miles from earth.

Scientists have learned one thing by many hundreds of years of investigations. God made a huge universe. We still do not know the size of this very big universe.

Buzz Aldrin on the Moon, 1969

CHAPTER 2: GOD MADE GLORIOUS SIGHTS

Our Creator God made one gigantic sun for us. He made all of the stars to shine brightly throughout the entire universe. They shine with immense power. The size of the universe and its powerful energy is truly spectacular. It is overwhelming to think about it. And all of this gives us only a tiny peak into the immense, all-powerful energy of God Himself. All praise be to God for His great, unlimited power and might!

Therefore David blessed the LORD before all the assembly; and David said:
"Blessed are You, LORD God of Israel, our Father, forever and ever.
Yours, O LORD, is the greatness,
The power and the glory,
The victory and the majesty;
For all that is in heaven and in earth is Yours;
Yours is the kingdom, O LORD,
And You are exalted as head over all.
Both riches and honor come from You,
And You reign over all.
In Your hand is power and might;
In Your hand it is to make great
And to give strength to all."
(1 Chronicles 29:10-12)

How Big is God's Universe?

This is a big question. The simple answer is that we don't know. We cannot see to the end of it, even with the best telescopes. The size of the universe is a mystery that only God knows. He Himself is infinitely more immense than the universe itself, and the universe He created is a testimony to His infinity. We are humbled. We are in awe of God, as we look in wonder at the sheer hugeness of His universe!

Scientists have guessed that the universe is 28 billion **light-years** from one end to the other. Some believe that the universe is 250 times larger than that, or about 7 trillion light-years from end to end. How big is a light-year? That's the distance light will travel in a year, and light travels very fast—186,000 miles per second. If you were to climb into a rocket and travel at 186,000 miles per second or 670,000,000 miles per hour, it would take you a year to travel one light-year.

Let's put this another way. Let us say that the whole earth is just one little dot. The first star (which would be our sun) would be 3,673 feet away from us.

A **galaxy** is a collection of billions of stars, tons of dust, and gases that hang

Milky Way Galaxy

together in one area of the universe by a force called **gravity**. We live in a galaxy called the **Milky Way**. Our sun is only one star in that Milky Way, out of 250 billion stars (or suns).

Our earth is pretty big—about 8,000 miles wide. But suppose the earth were just a little dot on this page. Our sun would be 3,673 feet away from us. But, the edge of the Milky Way Galaxy would be 1,320,000,000 miles away from us.

The sun is about 92,000,000 miles from the earth. If you were to climb into a rocket ship and travel at 70 miles per hour, it would take 150 years for you to arrive at the sun. Our best rocket ship can travel at 430,000 miles per hour. At this rate, you could get from New York City around the world to Beijing in one minute, or you could get to the moon in half an hour. But it would take you 210 hours in that rocket ship to make it to the sun. It would take 5,300 years to get to the next star. If you kept riding on the rocket ship, it would take 45 million years to get to the edge of the Milky Way Galaxy. That means, it is highly doubtful that humans will ever make it to another solar system.

CHAPTER 2: GOD MADE GLORIOUS SIGHTS

In the foreseeable future of this world, man will never explore any other planets outside of this solar system. Should the end of the world come within the next thousand years, humans will never make it to another solar system. God created the earth for us, and He did not intend for us to live anywhere else (Psalm 115:16).

Now, our Milky Way is only one galaxy in a very large universe of at least 100 billion galaxies. Riding in our rocket ship at 430,000 mph, it would take 42 trillion years to reach as far as we can see of the outer reaches of the universe.

How big is this universe in reference to little you? If a pilot were to fly in an airplane over your town, it would be very hard to see you. You would be a tiny dot in your front yard.

Your whole city is a tiny dot on the map of the country where you live. Your country is only a tiny piece of the earth.

The earth is only a tiny dot from the vantage point of the sun. The earth could not be seen from the sun (our closest star) without a telescope.

From the edge of the Milky Way Galaxy, our sun is only a tiny dot which could not be seen in the mass of stars in our solar system.

Our solar system is only a tiny dot in the universe, almost impossible to see without a telescope from another galaxy. Now, how big are you in this universe?

When I consider Your heavens, the work of Your fingers,
The moon and the stars, which You have ordained,
What is man that You are mindful of him,
And the son of man that You visit him?
(Psalm 8:3-4)

Where We Live

We live about 92,000,000 miles away from the sun in the Milky Way Galaxy. We live on the planet earth, which is one of eight planets in our solar system. All of the planets in our solar system revolve around the sun. We call this a **solar system**, because the sun is the center of it, and the sun provides heat and light for the planets.

The **planets** in our solar system that we know about are Mercury, Venus, Earth, Mars, Jupiter, Saturn, Uranus, and Neptune. Pluto is a Dwarf Planet, and so it is not included in the list of full-fledged planets. However, scientists believe there

GOD MADE THE WORLD

is one more gigantic planet somewhere far out in our solar system, called **Planet X**.

The planets which are closest to the sun are very hot, and the planets farther away from the sun are very cold. All planets rotate on an axis and revolve around the sun. The length of day depends on how long it takes for the planet to spin 360 degrees on an imaginary axis. The earth spins like a top, and people standing on the equator are spinning at 1000 miles per hour! Because God keeps the earth spinning at a constant rate, people don't even realize they are moving that fast. If it came to a sudden stop, then everybody, your pets, your house, and your car would be flung out into the atmosphere.

The length of year for each planet depends on how long it takes for it to revolve around the sun.

The location of the planets is measured by the distance of the earth to the sun. So Jupiter is over four times further from the sun than the earth is from the sun.

What would it be like to work for 700 hours, and then take a 700 hour rest at night on Mercury? You would be all tired out after working 700 hours in a row!

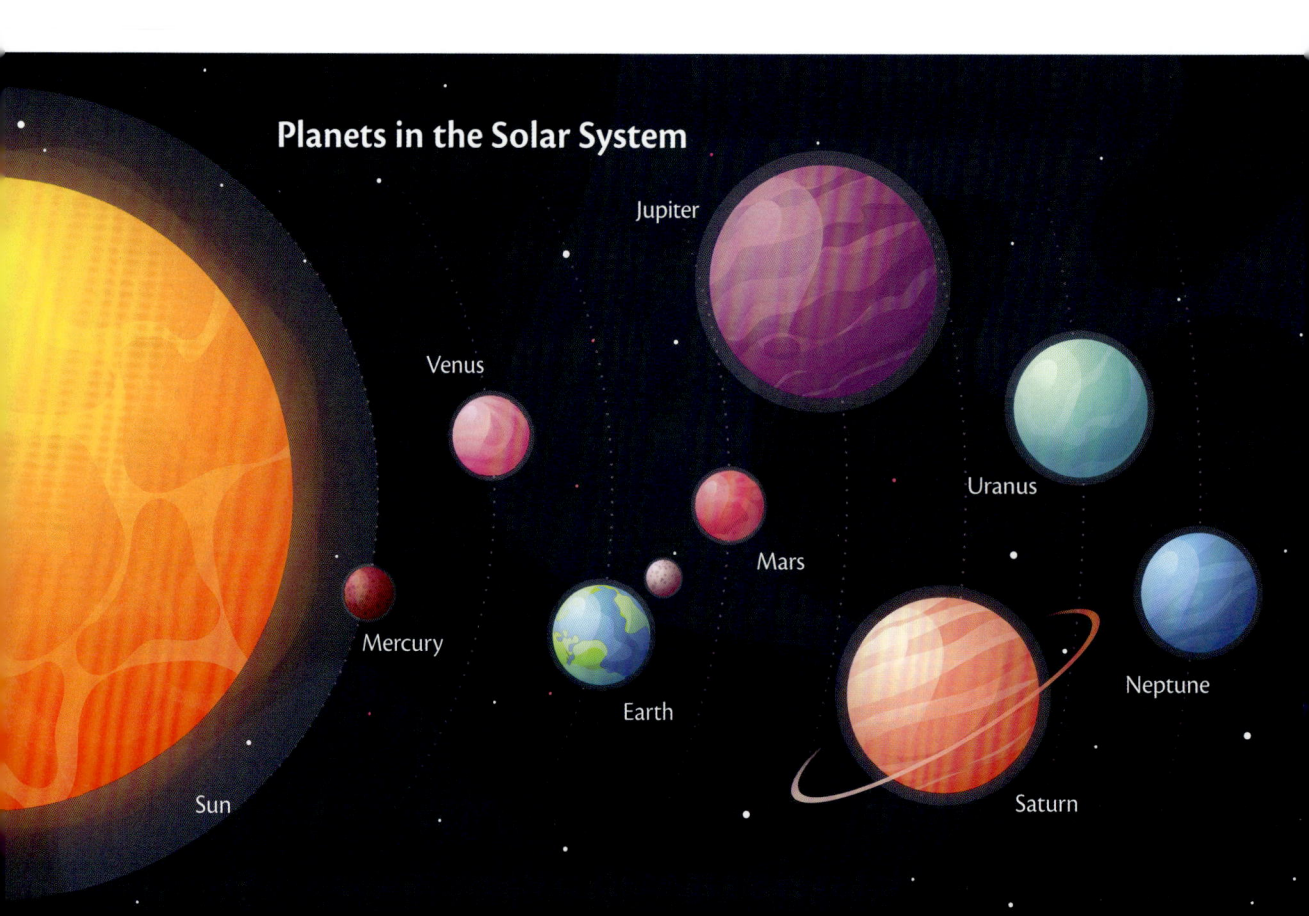

Planets in the Solar System

Sun • Mercury • Venus • Earth • Mars • Jupiter • Saturn • Uranus • Neptune

CHAPTER 2: GOD MADE GLORIOUS SIGHTS

Planet	Length of Day	Length of Year	Average Temperature	Proportionate Distance from Sun
Mercury	1,408 hrs	88 days	333 °F (167 °C)	.3
Venus	5,832 hrs	225 days	867 °F (464 °C)	.7
Earth	24 hrs	365 days	59 °F (15 °C)	1.0
Mars	25 hrs	687 days	-85 °F (-65 °C)	1.3
Jupiter	10 hrs	12 years	-166 °F (-110 °C)	4.4
Saturn	11 hrs	30 years	-220 °F (-140 °C)	9
Uranus	17 hrs	84 years	-319 °F (-195 °C)	19
Neptune	16 hrs	165 years	-328 °F (-200 °C)	45

Dwarf Planets	Length of Day	Length of Year	Average Temperature	Proportionate Distance from Sun
Ceres	9 hrs	4 years	-159 °F (-106 °C)	2.7
Pluto	153 hrs	248 years	-378 °F (-228 °C)	49
Haumea	4 hrs	285 years	-400 °F (-240 °C)	50+
Makemake	8 hrs	309 years	-405 °F (-243 °C)	50+
Eris	8 hrs	557 years	-414 °F (-248 °C)	50+

GOD MADE THE WORLD

Would you want to go through a 343 day fall and winter on Mars? The average temperature of 333 ºF (167 ºC) on Venus would be way too hot for human life. How would like to live in an oven at 333 ºF (167 ºC) every day?

There are so many different reasons why this planet is the perfect location for us. It's neither too hot or too cold for growing things. A full 32% of the earth's surface grows vegetation. There is sufficient water, neither too much or too little—plenty of room for billions of people to live and to grow food. The earth's surface includes plate tectonics that regulates the temperature for us. The earth is just the right size to maintain just enough atmosphere that it doesn't get too hot for us. There's a lot of debris in space, including asteroids which could get pulled into the earth's gravity and create huge explosions on the earth. But our "big brother" planet Jupiter collects a large amount of this debris for us, keeping us safe. In fact, a huge asteroid belt between Jupiter and earth contains about two million asteroids from 0.6 to 293 miles in diameter/width. A 60-mile-wide asteroid would destroy all life on earth.

God made a planet for us that would be just the right place for life. Let us praise God for His goodness to man!

> For thus says the LORD,
> Who created the heavens,
> Who is God, Who formed the earth and made it,
> Who has established it,
> Who did not create it in vain, Who formed it to be inhabited:
> "I am the LORD, and there is no other."
> (Isaiah 45:18)

Mars

The Sun's Nuclear Fusion Reaction

How Big and Powerful Are the Stars?

*For all the gods of the peoples are idols,
But the Lord made the heavens.
(Psalm 96:5)*

Our sun is an average-sized star, about 864,000 miles wide. That is much bigger than the earth, which is only about 8,000 miles wide. If the earth were the size of a golf ball, the sun would be the size of a large bus.

The sun burns as a gigantic nuclear energy plant. Hydrogen atoms are compressed and fused together in an explosion called **nuclear fusion.** In the process, helium is created. These gases heat up. Atoms break apart into charged particles turning the gas into plasma. The surface of the sun is about 10,850 ºF (6000 ºC), but temperatures sore to 15,000,000 ºC at its core. The hottest fires on earth burn at 5,400 ºF (3000 ºC). Imagine one trillion nuclear bombs exploding every second!

The sun is burning 8 billion tons of solar material every second. So it is getting smaller all the time, but there is a lot up there to burn. Don't be concerned that the sun will run out of fuel any time soon. Your fireplace runs out of logs after a few hours, and the fire goes out. But God has created the stars with a lot of fuel, so they can burn for a very long time. Scientists guess that the sun can burn for another 8 billion years.

The largest star found so far is the Canis Majoris. If the earth were the size of a golf ball, Canis Majoris would be the size of Mount Everest. Suppose that you were to walk around the whole earth, exploring every square mile. It would take 6,000 years for you to visit every square

Stars in the Milky Way Galaxy

mile of this earth. But if you were to explore every part of Canis Majoris by foot, it would take you 10,000,000,000,000,000 years to do it. And that's just one star. Astronomers are people who study the stars. These scientists estimate that there are 1,000,000,000,000,000,000,000,000 stars in the universe. It would take you a very long time to explore all of those stars, and the planets that revolve around them.

God made all of this. God is still infinitely bigger than that whole universe, and He is more powerful than all of the powerful stars that burn in the heavens.

Our God is infinitely bigger, stronger, and more glorious than everything He has made. No power comes close to being stronger than God. He is way, way more powerful than everything He has made.

God made the earth, the sun, the moon, and all of the stars. He is more powerful than all the things He has made. Sometimes, we think that we are very strong. But, compare God's power

CHAPTER 2: GOD MADE GLORIOUS SIGHTS

to man's strength. The strongest man on earth can lift 1,100 pounds up to his hips. That is about the weight of a 20 inch rock. The earth and the stars are much, much bigger than that! The strongest man in the world is very, very weak compared to God.

Praise the LORD!
For it is good to sing praises to our God;
For it is pleasant, and praise is beautiful...
He counts the number of the stars;
He calls them all by name.
Great is our LORD, and mighty in power;
His understanding is infinite.
(Psalm 147:1,4-5)

From Psalm 147 we learn that the Lord doesn't lose track of any of His stars. He calls them all by name. Imagine keeping track of 1,000,000,000,000,000,000,000,000 stars every day, all day long! This reveals God's intelligence, His attentiveness, and His power over all things. All we can say is, "God is amazing!" Indeed, He is worthy of our grandest thoughts and our loudest shouts of praise.

Can You See the Stars and Planets?

Looking up into the night sky, you can see a few thousand of stars and a few planets. The planets appear brighter and bigger because they are much closer to us. However, the light from the planets is not generated from the planets. Those lights are only the reflection of our sun upon these giant masses in space. So the planets appear to be light and bright, when you look at them in the night sky. The sun is only shining on the planets.

However, stars do emit light, because, like our sun, they are gigantic burning balls of fire in space. Unlike planets, the stars appear to twinkle. When starlight breaks through the layers of the earth's atmosphere, the light beams shift angles slightly. Second by second, the light appears to come from a slightly different location as it works its way through the atmosphere.

Sunlight Lightens Planets

GOD MADE THE WORLD

God's Fireworks

People like to use spectacular fireworks displays to celebrate holidays. Maybe you have seen these bright colors exploding in the air. These are little explosions that are produced by man. These firecrackers are sort of powerful. But God's explosions are far more spectacular and powerful.

Sometimes our Creator God will make stars to explode. When this happens, they produce more energy in a second than our sun would produce burning for 10 billion years. These are called **supernovas** or **hyper-novas**. If a hyper-nova were to explode in the Milky Way Galaxy close to where we live, night-time would appear like day for a while. The star explosion would create that much light, even though the star was thousands of light years away from us. The light would gradually fade over a couple of weeks.

These are more demonstrations of God's amazing power!

The shockwaves that come out of the explosion can create new stars. The remnant material of the old star spreads

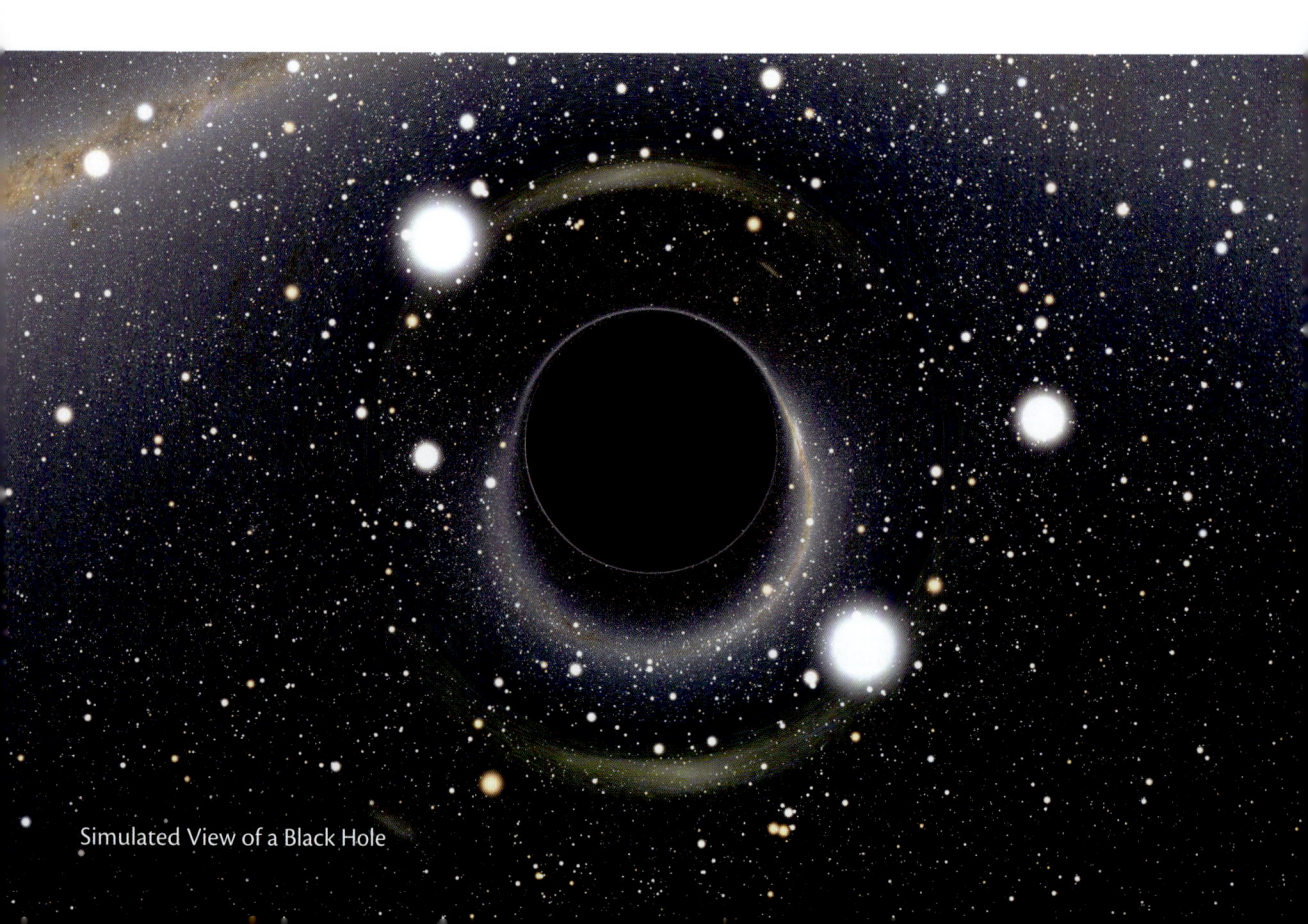

Simulated View of a Black Hole

Nebula

out, at first very quickly, almost at the speed of light. Using powerful telescopes, we can see the burning particles spreading out very far.

Sometimes the inner core of the old exploded star turns into a black hole. Imagine a huge giant taking the whole earth in his hands and scrunching and squeezing it all down into a very small marble. That's what it takes to make a black hole. This is called a star **implosion** as the star mass collapses in on itself. Because there is a huge amount of heavy mass in a very small area, a gigantic gravitational force is created. Like a humongous vacuum cleaner, the black holes will suck all the mass floating around and the light emanating from stars in the vicinity into its core. The black holes themselves cannot produce light. Nothing can escape from a black hole.

Nebulae

Then the LORD answered Job out of the whirlwind, and said:
"Where were you when I laid the foundations of the earth?
Tell Me, if you have understanding.
Who determined its measurements?
Surely you know!

GOD MADE THE WORLD

Or who stretched the line upon it?
To what were its foundations fastened?
Or who laid its cornerstone,
When the morning stars sang together,
And all the sons of God shouted for joy?
(Job 38:1, 5-7)

For thousands of years, man could not see much of the beauty of the universe. Only God could see the beauty of the Nebulae. Now, with telescopes we can see these amazing displays of God's fireworks in the sky!

The **nebula** is a cloud of dust and gas, coming out of an aging or exploded star. The most famous Nebulae include the Butterfly, the Intricate Crab, the Splendor of Orion, Helix, Double Helix, Red Cosmic Square, Stingray, Red Spider, Hourglass, Tarantula, Cat's Eye, Ant, and the Boomerang.

The largest nebula light show in the universe as far as we can tell is the Tarantula. This incredible display of God's wisdom, beauty, and power is 1,000 light years across. That would be about 6,000,000,000,000,000 miles across. How would you like to paint a beautiful mural that size? Surely, God does His artwork in a very big way! You can see this nebula without the aid of a telescope.

Hypergiant Stars

God is light, and in Him is no darkness at all. (1 John 1:5)

Our wondrous Creator also made **hypergiant stars**. These are huge stars which are just about ready to blow up. It might take a star thousands of years to prepare for an explosion. You might think of a star as a log burning in the fireplace. When it has burned the inside core up so that the outside burns hotter and faster,

Great Nebula Surrounding Eta Carinae

and the smoke particles blow off faster and faster. These hypergiant stars burn very hot and bright and as time goes by they lose mass faster and faster.

Truly, these lights declare the glory of God, and the holiness of God in the bright shining of an ineffable, blinding light. The Pistol Star shines at 10 million times the brightness of our own sun. You would never want to look into our sun directly. If you did so, you would probably go blind. What would happen if the Pistol Star took the place of our sun? Surely, the earth would burn up instantly.

Quasars

That you keep this commandment without spot, blameless until our LORD Jesus Christ's appearing, which He will manifest in His own time, He who is the blessed and only Potentate, the King of kings and LORD of Lords, who alone has immortality, dwelling in unapproachable light, whom no man has seen or can see, to whom be honor and everlasting power. Amen.
(1 Timothy 6:14-16)

Yet, there is still another much more powerful and brighter object than the hypergiant star shining in space. This is known as a **quasar**. Scientists can only guess what makes them shine so brightly.

Galaxies are often centered around a large black hole. Around the black hole is sometimes found a quasar. So far, astronomers have identified 200,000 of these spectacular celestial displays. Deep in the constellation of Virgo, you can see the quasar 3C-273. This tremendous energy source shines at a brightness four trillion times the brightness of the sun. This one light source shines brighter than one hundred galaxies the size of our Milky Way!

Imagine four trillion of our suns in the sky, or one hundred of our Milky Way galaxies stuffed into one quasar all shining down on you! That's a lot of energy and light! That would burn up the whole earth in a micro-second! Once again, we get a picture of God who is light. He creates all of these exceedingly bright lights in the heavens. And in Him is no darkness at all.

God came from Teman,
The Holy One from Mount Paran. Selah
His glory covered the heavens,
And the earth was full of His praise.
His brightness was like the light;
He had rays flashing from His hand,
And there His power was hidden.
(Habakkuk 3:3-4)

Extraterrestrial Cyclones

Our Creator God has placed storms on some of the other planets in our solar system. These are big storms, spectacular storms, very, very powerful storms. The planet Saturn sports a "Great White Spot" storm that erupts now and then and that is the size of the earth. Wind speeds run at 330 mph (530 kmph). Sheets of

Great White Spot on Saturn

CHAPTER 2: GOD MADE GLORIOUS SIGHTS

lightening about as wide as the earth itself flash almost continuously for a spectacular light show! Jupiter's storm appears as an eye. This storm is larger than the whole earth, and it has been going on for over 400 years. Compare that to storms on earth that only last one or two days.

The winds in Jupiter's great storm blow at 275 mph (430 kmph). What would happen to your house, if winds would blow at 275 miles per hour continuously? The hardest wind gust occurring during a storm on earth was clocked at 212 miles per hour (340 kmph) on August 30, 2008 in Cuba. The strongest tornado in history blew at 302 mph (486 kmph) in Bridge Creek Oklahoma. It destroyed over 8,000 homes and killed 36 people. It was the most dangerous tornado in recorded history. Now, imagine that tornado covering the whole world for 400 years. Everybody would be killed and every house would be flattened in ten seconds!

God's power is shown, and His glory is seen in these amazing storms! We are so thankful that He keeps them at a distance so that human life can be preserved here on earth.

Why does the Lord order a great storm on Jupiter or Saturn? Of course, this is a demonstration of His great power. These storms would destroy all of life on earth, if God were to bring them here.

Clouds of Jupiter

Why God Made the Sun and Moon

Then God said, "Let there be lights in the firmament of the heavens to divide the day from the night; and let them be for signs and seasons, and for days and years; and let them be for lights in the firmament of the heavens to give light on the earth;" and it was so. Then God made two great lights: the greater light to rule the day, and the lesser light to rule the night. He made the stars also. (Genesis 1:14-16)

GOD MADE THE WORLD

Our God made the sun, the moon, and the stars to shine His power and glory. But, here in Genesis 1 we find more purposes for this creation. The sun provides light by day, and the moon offers light by night.

God created the earth to revolve around the sun. He gave us a moon which travels around the earth. And He has the earth rotate or spin on its axis over a 24-hour period.

At any given moment, the sun is shining on half of our ball-like earth. The other half of the earth is facing away from the sun, and so it is dark on that side. Since the earth rotates on its axis over 24 hours, this provides about twelve hours of light and about twelve hours of darkness for most of the earth. When China and India are facing the sun, the United States and Mexico are facing away from the sun. When it is daytime in China and India, it will be night-time in Mexico and the United States.

The moon is an amazing creation of God to provide light at nighttime. The

Sun Illuminates Earth and Moon

moon is 400 times smaller than the sun (in diameter), but it is also 400 times closer. That makes the moon about the same size as the sun from our vantage point. Consider also that the moon stabilizes the earth's position. Without the moon, scientists believe the earth would flop around as far as 85 degrees (compared to its ordinary 22.1-24.5 degrees). There would be wildly varying climates everywhere around the world. Africa, Europe, North America, China, and elsewhere would turn into the North Pole for long periods of time.

The Purpose of Stars

The stars and the sun basically stay in the same positions. But, as the earth rotates and revolves around the sun, the location of stars (and familiar groups of stars) look like they are moving. If you were to look at the sky tonight and tomorrow night, your favorite star (or group of stars) would be in pretty much the same place at 8:00 pm. But, as the months go by these stars would change position from your point of view. That's because the earth is changing its position month by month as it revolves around our star, the sun.

Also, the stars seen by children living in Australia or Brazil (in the Southern Hemisphere) are different than the stars seen by children living in the Northern Hemisphere (in Canada or Japan). When children in the Southern Hemisphere are looking up into the sky, they are looking the opposite direction of those who are looking up in the Northern Hemisphere.

Certain familiar groups of stars are called **constellations**. If you live in America, you will see the following constellations throughout the year. These constellations form a map in the sky, and over the years they were very helpful for explorers and sea travelers.

Scorpius the Scorpion Constellation

They could figure out where they were by the location of the stars. If sailors were on the open seas, they could tell which way was north, by searching for the North Star. If the North Star was straight ahead, they knew that East was to the right, and West was to the left.

Constellations Seen in the Northern Hemisphere

Month	Constellation
December, January, February	Orion the Hunter
March, April, May	Leo the Lion
June, July, August	Scorpius the Scorpion
September, October, November	Aquarius the Water Carrier

The largest constellations seen in the Southern Hemisphere are Hydra, Virgo, Cetus, Eridanus, Aquarius, and Ophiuchus.

God Made Seasons

Our earth has an imaginary line running around the widest part of it called the **equator**. Although people living near

Leo the Lion Constellation

the equator experience rainy seasons and dry seasons, they do not have the long summer days or short winter days of those who live in Australia, England, China, or the United States. Our friends at the equator enjoy 12-hour days and 12-hour nights all year long.

The earth revolves around the sun on an **elliptical** (not quite round) path. The sunshine is not quite directed at the middle or the equator. Some parts of the year, we are closer to the sun than at other times, but that is not what gives us the seasons of spring, summer, fall, and winter. Otherwise those on the equator would also experience the seasons, and they do not.

The reason for the seasons is that the

CHAPTER 2: GOD MADE GLORIOUS SIGHTS

earth is tilted. What if you twisted your back, and you walked around the house slightly bent? About half the year, the northern part of the earth tilts away from the sun as it travels around the sun. That's winter time. The other half of the year, the northern part of the earth tilts towards the sun. That's summer time. Whichever hemisphere (north or south) is tilted toward the sun is the side which receives more direct heat and light.

If you live in the Northern Hemisphere, in the month of December, do you see that sun straight up in the sky at noon time? Of course not. In the month of June, you will see the sun almost straight up in the sky at noon time. Whichever hemisphere is tilted toward the sun receives more direct light and heat, and it gets more daylight hours. When it is summer time in the Northern Hemisphere, while the sun is high in the sky, the Southern Hemisphere is not getting as much sun. It will be winter in the Southern Hemisphere while it is summer in the Northern Hemisphere.

The **spring equinox** and the **fall equinox** happens when both hemispheres are receiving equal amounts of sunlight during the day and darkness during the night. A little boy living 1000 miles north

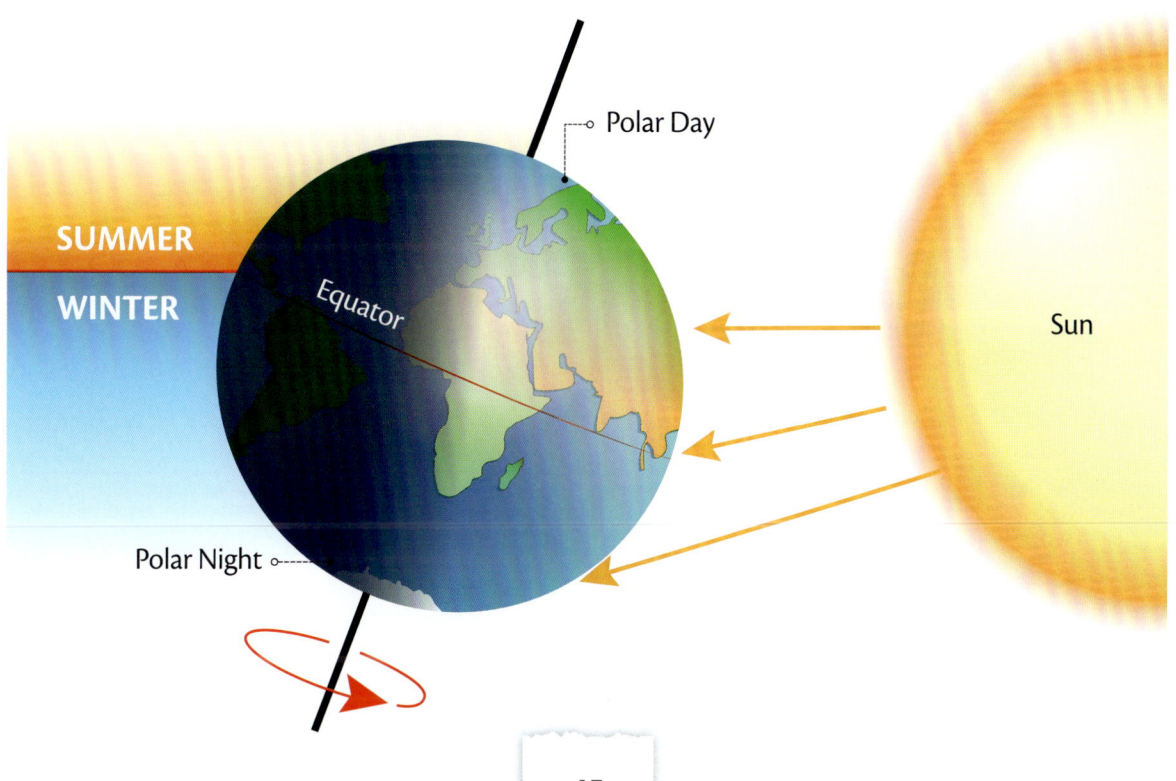

GOD MADE THE WORLD

of the equator receives equal hours of daylight and darkness as a little girl living 1000 miles south of the equator. The spring equinox for the child living in China is called the fall equinox for the child living in Australia. That's because March 21st is the first day of spring in China, but it is the first day of fall in Australia.

Seasons are a gift from God. He gives us warm seasons to help the crops in the fields to grow over about six months. Then the fields will rest for six months during the winter.

In ancient times, it wasn't too hard to figure out the length of a day. You would simply measure the time it took from sunrise on one day to sunrise the next day.

It was a little more difficult to figure out the length of a year. There are two accurate ways to determine the length of the year. First, you can figure out the number of days it takes to go from one Spring Equinox to the next Spring Equinox. The other way is to carefully analyze the location of a constellation

Sydney, Australia in the Springtime (in November)

The Great Wall of China in the Fall

Conditions near the Equator in the Democratic Republic of the Congo

at 8:00 pm, and then count the number of days, hours, and minutes it takes until the next time that constellation arrives at the same point. You have to be sure that you are standing in the same place and looking in the same direction when you take the reading.

This was the only way to establish calendars, schedules, dates, and times. God made the sun and the stars so that we could lay out times and seasons. This would help farmers know when to plant seeds, and it would help us all show up on time for church on Sunday.

> You visit the earth and water it,
> You greatly enrich it;
> The river of God is full of water;
> You provide their grain,
> For so You have prepared it.
> You water its ridges abundantly,
> You settle its furrows;
> You make it soft with showers,
> You bless its growth.
> You crown the year with Your goodness,
> And Your paths drip with abundance.
> (Psalm 65:9-11)

The Phases of the Moon

There are several things you need to know about the moon. Amazingly, God created the moon to rotate and revolve at the same time, such that everybody on Earth can only see one side of the moon. From Earth, we can never see what is known as the "dark side of the moon."

Throughout the 27.3 days that the moon is traveling around the earth, we watch the moon go through **phases**. Remember that the sun is always shining on the moon—one half of the moon. The **new moon** occurs when the moon is between the earth and the sun. You can see in the illustration that the sun is shining on the side of the moon that is facing away from us (which turns out to be what we call the "dark side."). The **full moon** occurs when the sun shines on the side of the moon that is facing us. When the moon is full, you know that the moon has rotated to the other side of the earth and is furthest away from the sun in its travel. At the first quarter and third quarter, we can only see half of the moon on which the sun is shining.

The moon moves eastward about 12 degrees each day as it circles the earth.

Eclipses

> Now from the sixth hour until the ninth hour there was darkness over all the land. And about the ninth hour Jesus cried out with a loud voice, saying, "Eli, Eli, lama sabachthani?" that is, "My God, My God, why have You forsaken Me?" (Matthew 27:45-46)

An eclipse happens when one celestial body moves in front of another, temporarily blocking our view of it. Because the moon and the sun are the two really big celestial bodies that we can see from earth, it's pretty obvious when the moon is obscured (lunar eclipse), or the sun is obscured (solar eclipse). When the sun is blocked by the moon in the middle of the day, suddenly everything gets darker.

Phases of the Moon

The **solar eclipse** happens when the moon passes in front of the sun.

The **lunar eclipse** happens when the moon passes through the earth's shadow.

These eclipses appear as total, annular, hybrid, or partial. A partial eclipse occurs when the moon covers only one part of the sun. An annular eclipse occurs when the moon covers the sun, but a ring of light still appears around the moon. A total eclipse produces a darkness (like night-time at noon) for about seven minutes in a 100 mile wide strip over a distance of about 10,000 miles. The hybrid is a combination of a total and an annual eclipse.

When Jesus died on the cross for our sins, darkness came upon the city of Jerusalem for three hours. This would have been 60 times more severe than any total eclipse that has ever occurred in the history of the world. Of course, God is able to perform a powerful miracle involving the largest celestial bodies in our solar system. However, the greatest miracle of all came three days later when God the Father raised Jesus from the dead.

The next total solar eclipses, visible from many nations around the world, will be:

GOD MADE THE WORLD

Date for Eclipse	Areas of the World Where Visible
April 8, 2024	Mexico and the United States
August 2, 2027	Spain and North Africa
July 22, 2028	Australia and New Zealand
November 25, 2030	South Africa
March 30, 2033	Russia, Alaska

Lunar Eclipse

Solar Eclipse

What Keeps Everything Going

Then Joshua spoke to the LORD in the day when the LORD delivered up the Amorites before the children of Israel, and he said in the sight of Israel:
"Sun, stand still over Gibeon;
And Moon, in the Valley of Aijalon."
So the sun stood still,
And the moon stopped,
Till the people had revenge
Upon their enemies."
Is this not written in the Book of Jasher? So the sun stood still in the midst of heaven, and did not hasten to go down for about a whole day. And there has been no day like that, before it or after it, that the LORD heeded the voice of a man; for the LORD fought for Israel. (Joshua 10:12-14)

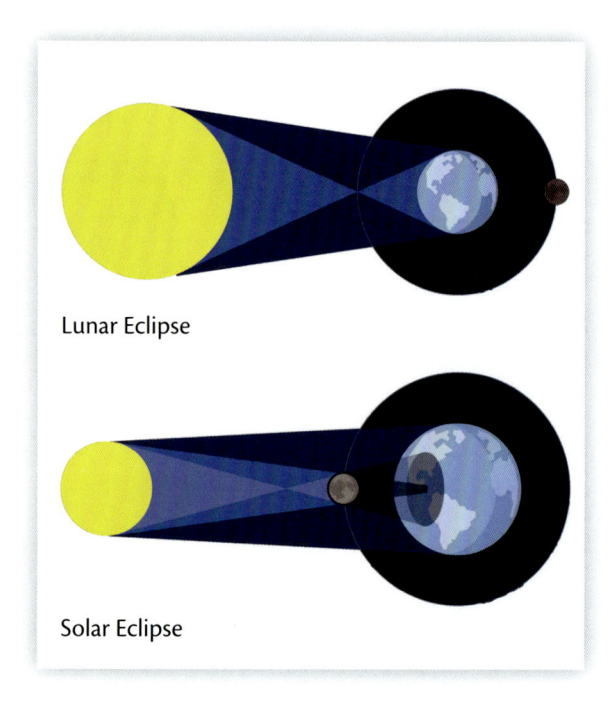

During this battle with the Amorites about 3,400 years ago, Joshua asked the Lord to keep the sun still in the sky for a whole day. Evidently, the earth kept still for at least twelve extra hours, because God made sure it would happen.

For the next 3,400 years the earth has gone back to spinning around every 24 hours. The moon keeps revolving around

CHAPTER 2: GOD MADE GLORIOUS SIGHTS

the earth every 27.3 days. The earth keeps revolving around the sun once a year. What is it that keeps everything moving? Of course, it is God who keeps everything moving. Everything holds together by the power of Jesus Christ, according to Colossians 1:17: "In Him all things consist."

It turns out that God uses regular, consistent forces in the universe to keep things going. God does not run the world in some chaotic or random fashion. He doesn't make summer eleven months long one year, and then make it one month long the next year. He doesn't make the earth turn in 24 hours one day, and then speed up the globe so that the next day is only 18 hours long. God is wise and kind to us, so we can plan our days and years. He gave Noah this promise in Genesis 8:

Then Noah built an altar to the LORD, and took of every clean animal and of every clean bird, and offered burnt offerings on the altar. And the LORD smelled a soothing aroma. Then the LORD said in His heart, "I will never again curse the ground for man's sake, although the imagination of man's heart is evil from his youth; nor will I again destroy every living thing as I have done.

"While the earth remains,
Seedtime and harvest,
Cold and heat,
Winter and summer,
And day and night
Shall not cease." (Genesis 8:20-22)

Gravity

God has made four mysterious, wonderful forces in nature. They are called:

1. Gravity
2. Electromagnetism
3. Strong Nuclear Force
4. Weak Nuclear Force

These are the forces that keep everything together. Without these forces, everything would fall apart. The earth and the moon would just float away, and there would be no order in the world. These are the forces that keep the universe running smoothly, always under the hand of God. In this chapter, we will only address the wonderful force of **gravity**.

When a child puts his glass of milk next to his elbow on the table, what usually happens next? Of course, his elbow hits the milk. Then, what happens? Does the glass of milk just sit there in the air right next

GOD MADE THE WORLD

Word of His power. This is what we read in Hebrews 1:

> God, who at various times and in various ways spoke in time past to the fathers by the prophets, has in these last days spoken to us by His Son, whom He has appointed heir of all things, through whom also He made the worlds; who being the brightness of His glory and the express image of His person, and upholding all things by the word of His power, when He had by Himself purged our sins, sat down at the right hand of the Majesty on high. (Hebrews 1:1-3)

the table? Oh no! The glass falls to the ground with a clatter, and milk spills everywhere! Why does the glass fall to the ground? Did somebody throw the glass on the ground? We usually say that the glass falls to the ground. Actually, there was a mysterious, unseen force that pulled it to the ground very quickly.

What if this force did not exist? Of course, everything would be floating in the air. If there was no force, nothing would fall to the earth. The waters would not flow in the riverbeds. All the dirt and rocks would float up into the air, and God's creation would be a big mess.

Praise God for His provision of this wonderful force called gravity! Indeed, it is the Lord Jesus Christ who holds the world together, and He does it by the

Every physical thing around you has a tiny force of gravity within it that pulls other things toward it. Even little things like pens, books, and cups have this mysterious force hidden inside of them. However, you notice these pens and books do not move towards each other. That is because that gravitational force is still way to small. However, really big masses, big things like the earth and the moon, have a big force which pull smaller things towards them.

The sun pulls the earth towards it. The earth pulls the moon towards itself.

CHAPTER 2: GOD MADE GLORIOUS SIGHTS

Imagine a gigantic rope pulling the moon towards the earth. Or think of a tether ball game where the ball moves in a circular path around the pole. But then, you ask, "Why doesn't the moon just smash into the earth? Or why doesn't the earth get sucked into the sun?" Of course, that would be very bad. The whole earth would get burned up and none of us would survive. At the beginning, the Lord gave the earth a push. He didn't push the earth towards the sun or away from the sun. He pushed the earth in a direction perpendicular to the sun, at about 67,100 miles per hour. Nothing would slow the earth down, because it was plowing through empty space. There was no wind to slow it down. However, the earth cannot keep going into space because that mysterious gravitational force pulls it toward the sun. So the earth keeps moving in a perpendicular direction, and the sun keeps pulling it towards itself. Gravity produces what is known as **centripetal force**, enabling the earth to continue its revolution around the sun year after year.

So the planets travel around the sun in this way. And the moon travels around the earth this way.

Tie a string to something heavy and

Planets in Orbit

GOD MADE THE WORLD

swing it around your head. You can feel the force of the heavy object trying to pull away from you. It takes a force to keep it from flying away. If you let it go, that heavy object would fly far away from you. Thanks be to God our Creator for gravity. This force keeps the earth from flying away into outer space. The force of gravity from the sun that is pulling on all the planets, keeps the planets from flying away into outer space. So, for over 6,000 years the planets continue to revolve around the sun, still moving at pretty much the same speed. As long as the planets keep revolving at the same speed that God gave them at the beginning, the length of a year on earth will remain about 365 days.

Everything pulls on everything else on earth and throughout space. Everything is pulling and pulling. Little objects have a tiny pull force. Big objects like the sun have a really big pull force. The further you are from the center of a big object, the weaker the pull will be on you.

How much gravitational pull do we feel from the sun, the moon, and the other planets? Standing on the earth, you can feel the earth pulling yourself down. When you drop a ball or anything, there is a force that yanks the ball down to the

Astronauts Floating in Space

CHAPTER 2: GOD MADE GLORIOUS SIGHTS

earth. This gravitation force produces an acceleration on the object dropped. Acceleration is what makes something go faster and faster.

If you fall off of a six-inch (15 cm) rock, you probably won't get hurt. That is because you are not going very fast when you hit the ground. It isn't a big deal to fall off of a six-inch rock. But, if you fell off of a rock that was six feet high, you would probably get hurt, because you are going much faster by the time you hit the ground. If you fell off of a 60 foot (18 m) high cliff, you would probably die, because you would be going very fast when you hit the ground. The gravitational force produces an acceleration on your body, that makes you go faster and faster as you fall.

We measure gravitational force in pounds/ounces or Newtons pound-force or ounce-force. The force that the gravity of the earth tugs on an apple when you drop it, it is only about 6.5 ounces (1.8 Newtons).

Some of the planets are smaller than ours, so you would weigh quite a bit less there. You would feel very light on Mars or Pluto, and you could jump much higher than you could on earth. Other planets like Jupiter are larger than ours, in which case you would weigh quite a bit more. This is because the gravitational force is stronger on larger heavenly bodies like Jupiter or Saturn. The force of the earth's gravity on a 100-pound child would be 100 pounds. The force of the earth's gravity on a 200-pound man would be 200 pounds.

You can't really feel the moon or the sun tugging on you. But the following table compares these tugs. How strong is the gravitation pull provided by the sun, the moon, and the other planets.

The Moon, the Sun, and the Big Planets Tug On You

Mass	Percent of Earth's Gravity	Force on 100 Pound Child
The Moon	0.35%	0.35 lbs
The Sun	0.06%	0.06 lbs
Venus	0.0001%	.0001 lbs
Jupiter	0.000005%	.000005 lbs

Now you can see that the moon has a much stronger pull on us than the sun, but it is not nearly as strong as the earth's pull. The moon is much smaller than the sun, but it is much closer to the earth.

GOD MADE THE WORLD

Remember, if you ride on the fastest rocket ship, you can get to the moon in 30 minutes. It would take you 210 hours to make it to the sun. So the gravitational force from the sun is much weaker than the moon, because it is so far away.

The planets are much smaller than the sun, and they are very far away. So we can hardly feel the tug from these planets on the earth.

Remember that the force of gravity from the earth keeps the waters sticking to the earth. These waters don't float away into space. However, the moon is also tugging on these waters. The moon's gravitational force is smaller than the earth's force of gravity. But, when the moon's gravitational force puts a little tug on the ocean waters, these waters actually move a little bit, this way and that way. The moon's gravitational force is what produces the tides in the oceans around the world.

The following table shows how much a boy would weigh on other planets or celestial bodies in our solar system. If he weighs 100 pounds (in earth weight), what would he weigh (in earth weight) on other planets?

Planet or Celestial Body	Weight of Boy Weighing 100 lbs (in earth pounds)
Sun	2,793 lbs
Moon	22 lbs
Mercury	38 lbs
Venus	91 lbs
Earth	100 lbs
Mars	38 lbs
Jupiter	234 lbs
Saturn	106 lbs
Uranus	92 lbs
Neptune	119 lbs
Pluto	9 lbs

If you were to travel to the moon and walk around on its surface, you would feel very light. A 100-pound child would weigh about 22 pounds on the moon.

Walking on the Moon (Elements Furnished by NASA)

The average person could jump ten feet into the air on the moon, and it would take about four seconds to come back down to the ground. A person would feel very heavy walking on the sun or on Jupiter. It would be very hard for him to jump up, as it is hard to jump when you are carrying 100 pounds of potatoes here on the earth. He would probably have a really hard time getting out of bed if he were living on the sun, because he would feel so heavy. Imagine feeling like you weigh almost 2,800 pounds! That would be like a small car sitting on your chest.

Now you can see the powerful force God made—the force called **gravity**.

What is this mysterious force called gravity? Nothing can shield you from this force. If you are anywhere near the earth, you will always fall back towards the earth, because the earth's gravity pulls you into it. Nobody really knows what produces this amazing force, except we know that the hand of God is behind it.

Isaac Newton may have been the greatest scientific mind of all time. He believed in God, the Creator. At the end of his life, he had to admit:

GOD MADE THE WORLD

I do not know what I may seem to the world, but as to myself, I seem to have been only like a boy playing on the seashore. . . now and then finding a smoother pebble or a prettier shell than ordinary, while the great ocean of truth lay all undiscovered before me.[1] (Isaac Newton)

What Does All of This Mean to You?

Are not two sparrows sold for a copper coin? And not one of them falls to the ground apart from your Father's will. But the very hairs of your head are all numbered. Do not fear therefore; you are of more value than many sparrows. (Matthew 10:29-31)

If Jesus holds the whole world together, and if God is all powerful, what does this mean for you? It means you must never be afraid of anything in the world. You must not fear men or sharks. You should only fear God.

As you look at the great and mighty mountains, oceans, and stars all around you, how should you think of the Creator? Let us put it very simply...God is far bigger and far stronger than everything He has

CHAPTER 2: GOD MADE GLORIOUS SIGHTS

made. As the children's song goes, "My God is so big, so strong and so mighty, there's nothing my God cannot do."

Yet, the most important lesson we take from all of this is that God can help us in our troubles. If God is that big and powerful, then He can help us when we are in trouble. He can save us from the bad things we do, and death itself.

He is always worthy to be praised. Scientists are often very proud when they learn something about God's universe. But God is far wiser than they are, for He created the universe! People love to praise other people for how smart they are, for their strength, or for the things they have accomplished. But God is worthy of all the praise, for He is the source of all good things. He created the human mind with the ability to learn things. He created this very complex world. He gives men their strength. He provides the great resources of light, energy, life, food, and water. All we do is receive these gifts and turn them into other forms that are useful to man. We are not very impressive, but God is endlessly amazing in His power, wisdom, and goodness! Amen.

GOD MADE THE WORLD

Pray
- Praise God for His truly awesome power! Praise Him for our powerful sun!
- Thank Him that the sun keeps spinning around in 24 hours every day. Thank Him for the seasons, for the tilting of the earth and the revolving of the earth around the sun.
- Thank Him for gravity. And praise Him for His sovereign power that keeps the whole world held together!
- Praise Him for the sun and the moon! Praise Him for the mighty stars, the great galaxies, the brightest quasars, and the gorgeous displays of Nebulae!
- Thank God that He pays attention to little us. What is man that God is mindful of us? Thank God that He sent His Son to visit us, and to save us from sin and death.

Sing
How Great Thou Art

O LORD my God, when I in awesome wonder,
Consider all the works Thy hands have made;
I see the stars, I hear the rolling thunder,
Thy power throughout the universe displayed.

Chorus:
Then sings my soul, my Savior God, to Thee,
How great Thou art! How great Thou art!
Then sings my soul, My Savior God, to Thee,
How great Thou art! How great Thou art!

When through the woods, and forest glades I wander,
And hear the birds sing sweetly in the trees.
When I look down, from lofty mountain grandeur
And hear the brook, and feel the gentle breeze.

CHAPTER 2: GOD MADE GLORIOUS SIGHTS

And when I think, that God, His Son not sparing;
Sent Him to die, I scarce can take it in;
That on the cross, my burden gladly bearing,
He bled and died to take away my sin.

When Christ shall come, with shout of acclamation,
And take me home, what joy shall fill my heart.
Then shall I bow, in humble adoration,
And there proclaim, "My God, how great Thou art!"

If you do not know the hymn, you may listen to a version of the hymn on the Internet, with supervision, and sing along with it.

Watch

To watch the recommended videos for this chapter, go to **generations.org/ GodMadeTheWorld** and scroll down until you find the video links for Chapter 2. Our editors have been careful to avoid films with references to evolution. However, we would still encourage parents or teachers to provide oversight for all internet usage. These videos may not give God the glory for His amazing creative work, so the student and parent/ teacher should respond to these insights with prayer and praise.

The Earth

Chapter 3
GOD MADE THE EARTH

> In the beginning God created the heavens and the earth. The earth was without form, and void; and darkness was on the face of the deep. And the Spirit of God was hovering over the face of the waters. (Genesis 1:1-2)

You live in the most important place in the whole universe—the earth! Why would we want to live anywhere else? God made the earth as the place where people would live. The earth is only a tiny speck in the whole universe, but it is God's special place. Out of His wisdom and goodness, at the beginning, He laid out the perfect place for us.

Lately, men have decided to explore space. Powerful governments spend lots of money on it. But, they do not want to glorify God. Unbelievers really want to prove that there is no God. Scientists want to find life on other planets. They think this will prove that the world came about by chance, without a Creator. Life is nowhere to be found everywhere else we look. It is only found on earth. Only God could have created this rare miracle of life!

Yet, these scientists still hope to find new places to live, such as on the planet Mars. These men and women want to prove that the universe developed by itself so they can ignore the Creator.[1]

They deny the Lord our God's creative work and His purposes for the universe. He created the earth as the one livable place for man. Then, He put the sun, the moon, and the stars in place for lights, for beauty, and for times and seasons. He did not intend for men to live any other place. We cannot live on the moon or on the other planets. We have everything we need to live here on the earth. There are no water sources on other planets that we know of. There is no air to breathe, no food, no trees for building things, and no safe place to live.

The United States has spent $1.17 trillion dollars exploring space. The best result of this effort has been to produce close up pictures of more of God's creation. We have verified that God has made very large planets to circle our sun. We have seen what they are made out of, and we have seen the glory of God's created work.

Before He created the sun, the moon, and the stars, God created our earth home. Then He created light, water, plants, trees, the heavenly bodies, and animals. Finally, on the sixth day He created man in His own image. And, He put Adam and Eve, the first man and woman, on the earth that was perfectly set up for them.

The Flood

*For God is my King from of old,
Working salvation in the midst of the earth...
You broke open the fountain and the flood;
You dried up mighty rivers.
The day is Yours, the night also is Yours;
You have prepared the light and the sun.
You have set all the borders of the earth;
You have made summer and winter.
(Psalm 74:12, 15-17)*

The earth we see today, with all of its mountains, lakes, rivers, deserts, and jungles is not the same place it was at creation. Two big events happened which changed the physical world where we live.

1. Man fell into sin.
2. God sent a worldwide flood to judge the wickedness of mankind on the earth.

Man's sin in the garden of Eden brought death into the world. Both men and animals die as a result of the fall. Then, the worldwide flood changed the whole shape of the earth's surface. The big continents formed. Huge mountains pushed up out of the waters. The oceans

CHAPTER 3: GOD MADE THE EARTH

formed around the world, and Antarctica and the Arctic iced over.

Before the flood, there were forests and dinosaurs in Antarctica.[2] Recently, hyena fossils have been found in the northern Arctic regions. But today, hyenas wander around near the equator—in the African savannah. Today, the average temperature in Antarctica is -127°F (-83°C), and hardly any rain or snow falls there. How did a tropical jungle turn into a freezing desert?

The only answer is that God brought a tremendous change on the earth with the worldwide flood. When man fell into sin, God cursed the ground. He changed the natural laws of the world. It could have been much warmer in the Arctic and Antarctica before the flood. The Genesis record tells us that man lived much longer, and so did the animals. But after the flood, the world would run by different patterns. The Lord God also established seasons. The world before the flood and before the fall of man was very different.

Antarctica

GOD MADE THE WORLD

And Adam lived one hundred and thirty years, and begot a son in his own likeness, after his image, and named him Seth. After he begot Seth, the days of Adam were eight hundred years; and he had sons and daughters. So all the days that Adam lived were nine hundred and thirty years; and he died. (Genesis 5:3-5)

Noah's Ark

Journey to the Center of the Earth

Some children like to dig holes in the backyard, hoping that they can reach to the other side of the world. If they could dig one foot of dirt every day for 116,000 years, they would make it all the way through to the other side of the world. Our globe is a little bit oblong, with a 7,926-mile distance all the way through at the equator, and 7,900 miles all the way through from the north to the south pole.

However, the deeper you would dig, the hotter it would get. At the core of the earth, scientists believe the temperature is about 10,000°F (or 5,540°C). Also, it would become increasingly hard to dig, because you would come against solid iron in the core. Because the iron core is under tremendous pressure, it remains solid. The iron does not melt at high pressures like this, even though it is extremely hot.

But the day of the LORD will come as a thief in the night, in which the heavens will pass away with a great noise, and the elements will melt with fervent heat; both the earth and the works that are in it will be burned up. Therefore, since all these things will be dissolved, what manner of persons ought you to be in holy conduct and godliness. . . (2 Peter 3:10-11)

CHAPTER 3: GOD MADE THE EARTH

about 1,800 miles thick (2,900 km), and it would take you about 26,000 years to dig through it if you could handle the heat.

3. **The Core** is about 760 miles thick (1,220 km), and it would take you about 11,000 years to dig through.

The Lord Jesus Christ made a very big world for us. This is another insight into His great power!

All things were made by [Jesus], and without Him was nothing made that was made. (John 1:3)

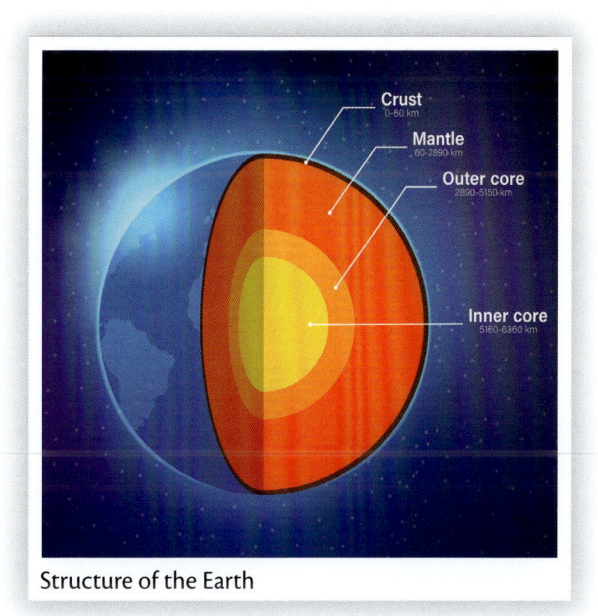

The earth will burn up at the end of time. This will happen at the day of judgment. Then God will create a new heaven and a new earth for us. But for now, God keeps the surface of the earth cool.

To get through to the other side of the earth, you would have to dig through three layers of the earth.

1. **The Crust** (where we live). This involves 4 - 28 miles (7-45 km) of digging. It would take you about 405 years to dig through the crust.

2. **The Mantle**. This part of the earth is made up of very hot rock. It is under high pressure which can cause it to shift around a bit. The mantle is

Structure of the Earth

67

GOD MADE THE WORLD

Ring of Fire

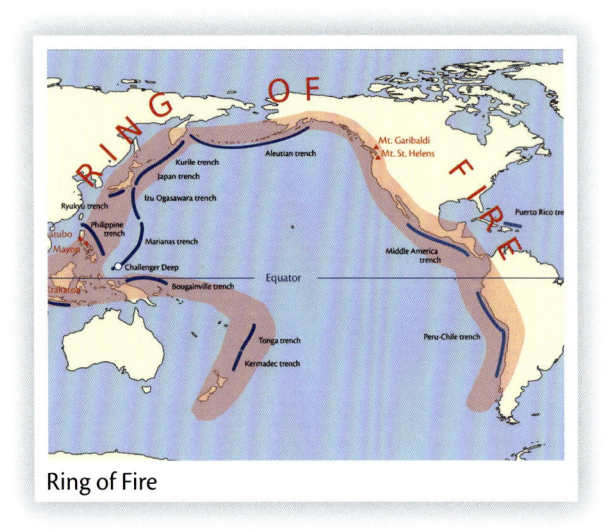

The Ring of Fire

At the time of the worldwide flood, God broke up the earth's crust, which included one very big crack. It is called the **Ring of Fire**. This is the source of many natural catastrophes that visit the world almost every year. About 75% of the earth's volcanoes and 90% of the earth's earthquakes happen along this crack.

Some creation scientists believe that this crack developed when the "great deep opened up" as we read in Genesis 7. When this crack opened up, a subterranean ocean

Mount Bromo, East Java, Indonesia

CHAPTER 3: GOD MADE THE EARTH

of water, sand, and rock exploded onto the earth and up into the atmosphere. By God's order, this crack opened up exactly on the other side of the world from where Noah and his family would have been floating about in the ark.

This Ring of Fire begins near New Zealand, and runs in a horseshoe pattern north through the Philippines. It continues along the coast of China and runs near Japan. Then it comes back down through Alaska, and the western coast of the United States, all the way down to the tip of South America on the west side.

On the following page is a list of the worst earthquakes of the last twenty years. All occurred on the Pacific Ring of Fire.

The three deadliest earthquakes in all recorded history, occurred in China in 1556, 1920, and 1976, taking 240,000-830,000 lives each. The most powerful earthquake ever recorded reached 9.5 on the Richter scale.

Scientists have developed a way to measure the severity of earthquakes on **seismographs**. The measurement is made on what is called the **Richter scale**. This is a measurement of the strength of earth movement, using a logarithmic scale. This means that a 6.0 earthquake is ten times more intense than a 5.0 earthquake. And, a 7.0 earthquake is ten times more intense than a 6.0 earthquake. Usually, a 7.0-8.0 earthquake would kill between 10 and 1,000 people in a city. An earthquake over 8.0 would kill between 1,000 and 100,000 people in a city.

These devastating earthquakes are reminders to the world that God is to be feared. Jesus warned the people facing mass disasters during His day, of God's judgment. He told them that everybody needs to take a lesson from these. You should repent of your sin before God, or you too will perish.

There were present at that season some who told Him about the Galileans whose blood Pilate had mingled with their sacrifices. And Jesus answered and said to them, "Do you suppose that these Galileans were worse sinners than all other Galileans, because they suffered such things? I tell you, no; but unless you repent you will all likewise perish. Or those eighteen on whom the tower in Siloam fell and killed them, do you think that they were worse sinners than all other men who dwelt in Jerusalem? I tell you, no; but unless you repent you will all likewise perish." (Luke 13:1-5)

Deadliest Earthquakes in History

Year	Magnitude	Death Toll	Location	Date and Title
2004	9.2	227,898	Indonesia	2004 Indian Ocean earthquake and tsunami
2008	7.9	87,587	China	2008 Sichuan earthquake
2011	9.1	20,896	Japan	2011 Tōhoku earthquake and tsunami
1999	7.7	2,415	Taiwan	1999 Jiji earthquake
2005	8.6	1,300	Indonesia	2005 Nias–Simeulue earthquake
2016	7.8	676	Ecuador	2016 Ecuador earthquake
2010	8.8	525	Chile	2010 Chile earthquake
2009	8.1	192	Samoa	2009 Samoa earthquake and tsunami
2000	7.9	103	Sumatra	2000 Enggano earthquake
2001	8.4	100	Peru	2001 southern Peru earthquake
2017	8.2	98	Mexico	2017 Chiapas earthquake
2007	8.4	23	Indonesia	2007 Sumatra earthquakes
2015	8.3	14	Chile	2015 Illapel earthquake

CHAPTER 3: GOD MADE THE EARTH

Seismograph

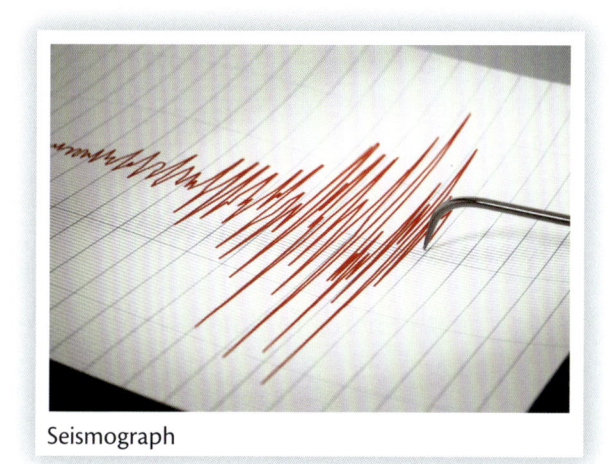

China, a nation controlled by communists, Japan controlled by Shintoism, and Indonesia (the largest Muslim-populated country in the world), need to heed the voice of warning from Jesus. They should repent of their sin and bow the knee to the Lord Jesus Christ, or they will perish. This goes for every nation in the world, including those who have turned away from Christ.

The Cause of Earthquakes

The LORD is slow to anger and great in power,
And will not at all acquit the wicked.
The LORD has His way
In the whirlwind and in the storm. . .
The mountains quake before Him,
The hills melt,
And the earth heaves at His presence,
Yes, the world and all who dwell in it.
Who can stand before His indignation?

2015 Illapel Earthquake Aftermath

GOD MADE THE WORLD

Tectonic Plates

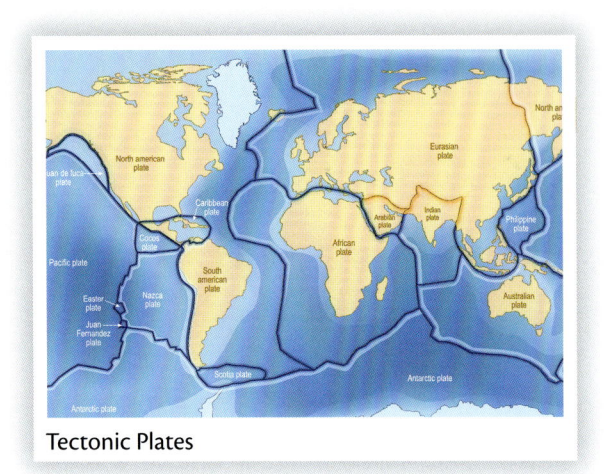

And who can endure the fierceness of His anger?
His fury is poured out like fire,
And the rocks are thrown down by Him.
The LORD is good,
A stronghold in the day of trouble;
And He knows those who trust in Him.
(Nahum 1:3,5-7)

The earth's crust is made up of large chunks of cracked plates, called **tectonic plates**. As God has ordered things to happen, these plates are always moving a little bit. Most earthquakes are caused when the plates rub up against each other. This will jerk the whole surface of the earth's crust in the area—an earthquake.

The Ring of Fire contains a great deal of hot magma, which adds to the instability for the plates sitting on the magma. This causes a lot of earthquakes, and volcanoes that break out in this area of the world.

The **epicenter** of an earthquake is the geographical location above where the biggest crash of tectonic plates takes place.

It's really hard to know when an earthquake will strike. Sometimes, seismologists can pick up little shocks on their seismographs. But they don't know whether these little shocks will turn into a big earthquake. Scientists try to guess when they will strike, but so far only toads have figured it out. The Journal of Zoology published a study in 2010 that discovered 96% of male toads in the region of the L'Aquila earthquake in

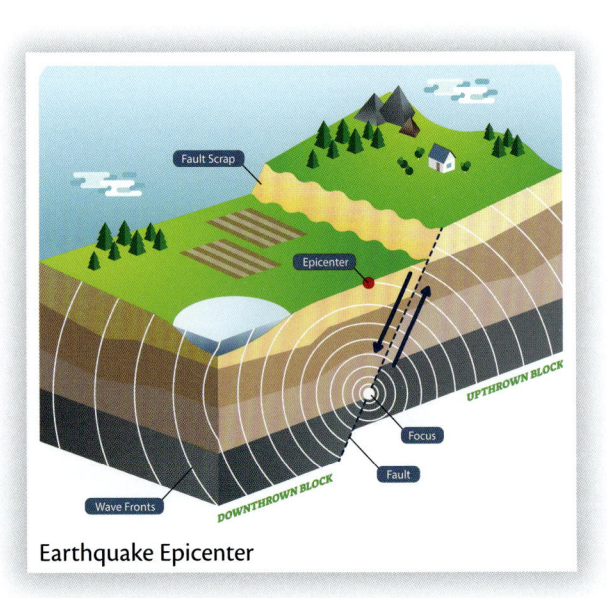

Earthquake Epicenter

Aftermath of the 2004 Tsunami in Indonesia

Italy abandoned the area five days before the earthquake hit.[3]

Actually, there is no place on earth that is safe from natural disasters. An earthquake can happen at any time. Only God knows when the next disaster will happen, because He has purposed it.

If a trumpet is blown in a city, will not the people be afraid?
If there is calamity in a city, will not the LORD have done it? (Amos 3:6)

The Tsunami or Tidal Wave

O LORD God of hosts,
Who is mighty like You, O LORD?
Your faithfulness also surrounds You.
You rule the raging of the sea;
When its waves rise, You still them.
(Psalm 89:8-9)

Earthquakes and volcanoes can produce huge waves called **tsunamis**, a Japanese word meaning "harbor wave."

GOD MADE THE WORLD

When an earthquake creates a disturbance on the ocean floor, this can result in a 10 to 100 foot wave traveling across the ocean into land.

Tsunamis can rush on to the shoreline at 600 miles per hour. What was probably the most powerful and largest tsunami in history hit Lituya Bay, in Alaska on July 9, 1958. The wave reached as high as 1720 feet (524 m), stripping a mountain of trees and vegetation.

The most tragic tsunami in history occurred on December 26, 2004. A gigantic, 110-foot (50 m) high wave crashed into Sumatra island in Indonesia. Over 200,000 people died. This wave was created by a huge, 9.1 level earthquake.

Volcanoes

May the glory of the LORD endure forever;
May the LORD rejoice in His works.
He looks on the earth, and it trembles;
He touches the hills, and they smoke.
I will sing to the LORD as long as I live;
I will sing praise to my God while I have my being. (Psalm 104:31-33)

Tungurahua Volcano Erupts in Ecuador

Mount Rainier, Washington, USA

The strongest energy man has ever been able to produce at one time was the nuclear bomb. The Soviet Tsar Bomba was detonated on an island in the north Arctic area in 1961. It produced the energy of 58 million tons of TNT. You could see the flash from the explosion 620 miles away (1000 km).

Nothing on earth compares to the power of God's volcanoes, however. The Tambora eruption of 1815 dwarfed all other eruptions since then. This single volcano produced more energy than the 10,000 volcanoes that followed. It would have been almost 15 times the energy of the Tsar Bomba. The explosion could be heard 1,600 miles away (2,400 km), and an estimated 80,000 people died from this very, very powerful natural disaster.

We have already seen how God's stars, His nuclear reactors in the sky, are trillions of times more powerful than what man produces. Yet we are just as impressed by God's volcanic explosions observed here on earth.

These are just little glimpses into the

Mount Vesuvius and the Ruins of Pompeii

power of God. Let us be amazed at His infinite might, and let us be reverent and worshipful as we think of His judgments in the earth.

"O LORD God of our fathers, are You not God in heaven, and do You not rule over all the kingdoms of the nations, and in Your hand is there not power and might, so that no one is able to withstand You?" (2 Chronicles 20:6)

Most volcanoes are caused by the same thing that makes earthquakes. In the case of volcanoes, fluctuating tectonic plates run into each other, and then one of the plates slides under the other. As the lower plate melts into magma, temperature and pressure increases. Gases that form inside the newly created magma push hard, trying to escape through the crust of the earth. But, where will the explosion of hot magma occur? Since the worldwide flood, some mountains are known to be volcanic. Mountains often sit on top of where the tectonic plates are stacked. They provide the vents for the eruption of gases and magma.

CHAPTER 3: GOD MADE THE EARTH

It's hard to know which volcano is going to blow next. But seismologists (people who study earthquakes) and volcanologists (people who study volcanoes) closely monitor areas where volcanic activity has been recorded in the past. Once again, most of the most active volcanic areas are located on the Pacific Ring of Fire.

- Mount Rainier—Washington State
- Mount Shasta—California
- Mount Hood—Oregon
- Mount Merapi—Central Java, Indonesia
- Mount Baker—Washington State
- Mount Mauna Loa—Hawaii
- Mount Teide—Spain
- Mount Galeras—Columbia
- Mount Nyiragongo—Congo
- Mount Unzen—Kyushu, Japan
- Mount Colima—Mexico
- Mount Hekla—Iceland

Two of the most disastrous volcanoes of all time occurred in ancient Greece and Rome. Somewhere around 1700 BC, Mount Santorini exploded with the force of hundreds of atomic bombs, destroying the Minoan civilization. Then, Mount Vesuvius in Rome erupted 79 A.D., destroying the most wicked city in the Roman Empire at the time—Pompeii.

Preparing for Earthquakes and Volcanoes

Behold, the eye of the LORD is on
Those who fear Him,
On those who hope in His mercy,
To deliver their soul from death,
And to keep them alive in famine.
(Psalm 33:18-19)

The most necessary lesson, and the first lesson if we will be prepared for all disasters and the judgments of Almighty God is to fear Him. Every child, every man and woman must study the fear of

Highway After Earthquake

God. He is to be feared, because He is God. He is over all. He is all powerful. He is the one who brings His judgments on the earth. Let us fear God, first of all. That means we are to look up to Him. We must ask forgiveness of God, in Jesus' name, when we sin against Him. We must think of God first, and humble ourselves before Him always.

Earthquakes and volcanoes are regular events in this fallen, sinful world. There are 20,000 earthquakes a year, but only about twenty that result in loss of life. Anywhere between 600 and 300,000 people die every year because of earthquakes. Only about 50-70 volcanoes erupt in a given year.

The earthquake itself doesn't really hurt anybody. If you were standing out in a field, you probably wouldn't be hurt by the earthquake, no matter how severe it was. However, earthquakes shake buildings hard, and things fall. Gas lines break. Fires are ignited. Automobiles sometimes crash into things. Most people are killed when parts of buildings fall on them, or they are caught in burning houses.

When you feel an earthquake start, remember that it will probably continue for 30-40 seconds. The largest earthquakes can go on for five minutes. Aftershocks might continue for hours afterwards. The ground under you will feel like it is swaying back and forth. Here's what you should do. Always stay calm. Pray. Trust in God. Remember He is in control of everything.

1. If you are indoors, drop to the ground, and try to take cover under a very sturdy table. Or you might crouch on the inside corner of an inner room. Stay away from windows and doors. Cover your head in your arms. Be sure to stay away from brick fireplaces, bookcases, and large cabinets that might fall on you.

2. If you are outside, stay away from buildings, electric poles and lines, and trees.

3. If you are driving in the car, pull over. Stay away from overpasses, and keep off all bridges. They might collapse. Stay in the vehicle, unless you might be hit by another car.

4. If you happen to find yourself in a large multi-storied building, remain on the same floor. Stay away from outside walls and all windows. And, do not use the elevator.

How Do We Best Prepare for Earthquakes Before They Happen?

"Therefore whoever hears these sayings of Mine, and does them, I will liken him to a wise man who built his house on the rock: and the rain descended, the floods came, and the winds blew and beat on that house; and it did not fall, for it was founded on the rock. But everyone who hears these sayings of Mine, and does not do them, will be like a foolish man who built his house on the sand: and the rain descended, the floods came, and the winds blew and beat on that house; and it fell. And great was its fall." (Matthew 7:24-27)

The biggest lesson to learn about disaster preparedness according to Jesus, is to pay attention to His Words. Respect His truth above everything else you read. And do the things that He commands.

These verses also contain a good hint for preparing for earthquakes. Build your house on the rock. The biggest mistake that people make is to build homes such that they would collapse when an

1980 Eruption of Mt. St. Helens

earthquake begins to shake the building. So the most critical thing a family can do when it comes to earthquake readiness is to be sure their home is well built, according to the following basic rules.

1. Houses should be built on rock or hard soil—not on sand and muddy soil.

2. Foundations and concrete walls should be reinforced with high ductility (E-class, 60-100 ksi) steel rebar. If you don't have steel running through your concrete, it can crumble easily during an earthquake.

3. Sometimes connecting buildings can prevent toppling.

How to Be Safe When You Are Near a Volcano

When Mount St. Helens in Washington State erupted at 8:42 am, Sunday morning, May 18th, 1980, it was like an explosion of 50 nuclear bombs. A 1300-foot landslide crashed into Spirit Lake. The initial blast torched everything on the mountain's north side, flattening an area about 15 miles by 10 miles. The ash continued to pour out of the volcano for the next nine hours. Fifty-seven people died, and a few escaped.

There are several things to beware of during a volcanic eruption.

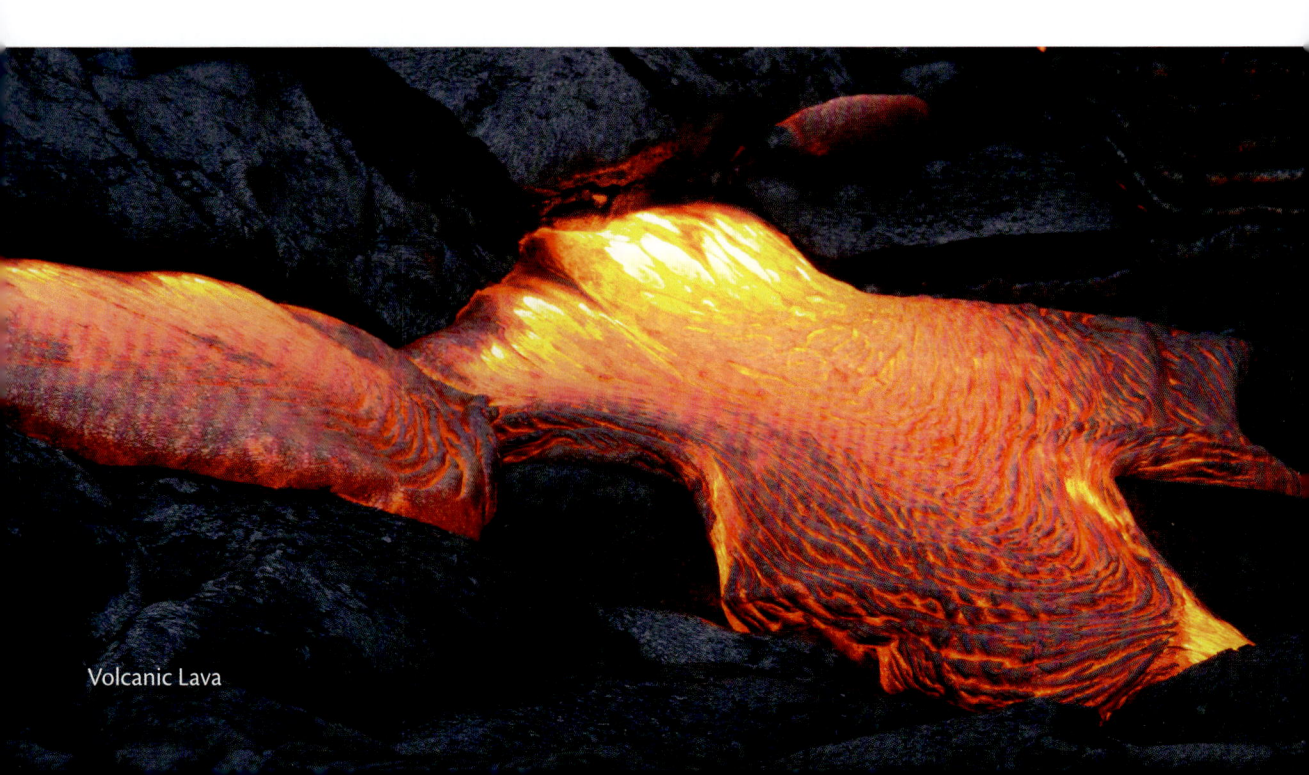

Volcanic Lava

CHAPTER 3: GOD MADE THE EARTH

1. **Lava flow.** You want to stay away from lava flow, because it is very hot. The flow can move at 10-40 miles per hour (15-60 km/hr). If the roads are clear going in the opposite direction of the volcano, an automobile should be able to outrun the lava flow. Do not attempt to cross over a lava flow. It would melt your tires or tennis shoes, and you would probably get stuck. Lava will tend to flow in the valleys. If you are on foot, climb up on ridges where the lava is less likely to flow.

2. **Ash.** If clouds of ash are falling around you, seek cover in a vehicle or building. Use a mask or a towel over your mouth to avoid breathing in the ash. If you are driving, try to avoid areas where there is a lot of ash accumulation.

3. **Flying Debris.** If you are fairly close to the volcano eruption, you may encounter flying rock and debris. Shield yourself behind or under large rocks, caves, vehicles, trees, or anything that might protect you.

Whatever natural disaster you might be encountering, tune into the radio to get updates on what is happening. Check the weather stations apps on your mobile phone. And obey the evacuation orders, and all other orders provided by the local governments.

If you are still at home here are a few other tips to follow:

1. Collect as much fresh water in the bathtub or other containers as you can.

2. Bring your pets indoors.

3. Wear goggles, safety glasses, and a mask or towel over your mouth if you go outside.

4. Watch for fire that might start up somewhere near the house, and put it out if you can.

5. Turn off your electricity and your gas, at least temporarily. You might have a professional check out your gas, before you turn it back on.

Mountains

LORD, You have been our dwelling place in all generations.
Before the mountains were brought forth,
Or ever You had formed the earth and the world,
Even from everlasting to everlasting, You are God. (Psalm 90:1-2)

GOD MADE THE WORLD

God made the mountains. These verses point out that the mountains we see around us were formed after the world was made. The shape of the earth we see today, was pretty much formed during and directly after the worldwide flood. Nobody has seen new mountains forming since then.

Mountains are huge. They appear immovable. There is a permanence to them. For thousands of years, man has looked up and seen the same mountains that their great, great, great, great grandparents looked at. Yet, God is far more permanent than the mountains.

Long before the mountains were made, before the earth was made, before angels were created, God was there. He has always been, and He will always be. So, therefore we can find salvation from death in Him. He can always provide us with protection from evil. His strength, His wisdom, and His goodness does not go away or diminish. We can always trust in Him.

Man might cut down a forest. He can dam up a river, and create lakes, but he cannot remove a mountain. A nuclear bomb cannot remove Mount Everest. Mauna Loa in Hawaii is the largest

Himalayan Mountains, Nepal

mountain (volume wise) in the world, with 18,000 cubic miles of mass. It would take 10,000 of the largest bulldozers in the world 1,000 years to remove this mountain.

"So Jesus said to them, 'If you have faith as a mustard seed, you will say to this mountain, "Move from here to there," and it will move; and nothing will be impossible for you'" (Matthew 17:20). There are harder things in life than moving mountains. We are sometimes challenged by very serious spiritual problems. We must respond to these problems by faith in God. It is by faith we can overcome our big problems.

How Mountains Were Made

In the six hundredth year of Noah's life, in the second month, the seventeenth day of the month, on that day all the fountains of the great deep were broken up, and the windows of heaven were opened. And the rain was on the earth forty days and forty nights. (Genesis 7:11-12)

Some mountains were formed when tectonic plates collided and pushed the earth's crust upwards. Others formed by volcanic activity. This happened mostly at the time of the worldwide flood. Remember that the deep parts of the earth broke up during the flood. This broke loose the gigantic tectonic plates, allowing for the magma under the earth to pour out onto the earth's surface. Scientists who do not believe in God think that

Paria River Valley, Utah, USA

Mount Everest, Nepal

GOD MADE THE WORLD

Igneous Rock

Sedimentary Rock

the mountains formed accidentally over millions of years.

There are three types of rock in the earth. **Igneous rock** is made of magma flow. This magma flow, you may remember, is the melted mantle of the earth. It is made mostly of **silica**. **Sedimentary rocks** are particles of sandstone and limestone compacted together by the flood waters. Layers of sediment formed as the flood repeatedly plowed mud and sand over the hills and valleys. The tides created by the moon probably helped the process along. These rocks are made of sandstone and limestone and some are thirty miles thick! Sedimentary deposits in Utah were carried all the way from Canada and the Eastern United States to make these hills. It is clear that these sedimentary layers were still wet when they were laid down by the flood waters in Utah. The Elbe Mountain range in southeastern Germany is also made of sandstone. **Metamorphic rocks** are made of igneous or sedimentary rocks, formed by extreme heating—which changes the shape of the rock.

The worldwide flood left the world with huge mountains and deep valleys. Study the table which lists the tallest mountains in the world.

Actually, the tallest mountain in the world is Mauna Kea on the Big Island of Hawaii. Its base lies deep in the ocean. If you were to measure it from its base, the mountain stands 33,474 feet high.

CHAPTER 3: GOD MADE THE EARTH

The Tallest Mountains in the World

Mountain	Height	How Long to Climb
Mt. Everest, Nepal Tallest Mountain in the World	29,029 ft (8,848 m)	60 days
K2, Pakistan Second Highest Mountain	28,251 (8,611 m)	60 days
Mt. Aconcagua, Argentina Tallest Mountain in S. America	22,841 ft (6,962 m)	15 days
Mt. Denali, Alaska Tallest Mountain in N. America	19,685 ft (6,000 m)	12 hours
Mt. Kilimanjaro, Tanzania Tallest Mountain in Africa	19,341 ft (5,895 m)	12 hours

Measured from sea level, however, the world's 30 tallest mountains in the world are in the Himalaya Range.

The most dangerous mountain in the world to climb is Mt. Annapurna in Nepal. About 40% of the people who have attempted the climb have died trying. Only about 90 people have succeeded, since the mountain was first conquered by man in 1950. K2 in Pakistan is the second most dangerous, with a fatality rate of about 20%.

Canyons and Rivers

Once the mountains started pushing up during the great flood, the waters began running off into the lowest spots on the globe into the ocean beds. Usually when huge amounts of waters run off, all the water creates deep river beds or **canyons**. That's what happened during the mighty flood which our Lord God brought upon the earth. Where did all the flood waters collected over Australia, Africa, North America, and Asia (Nepal and Tibet) go? The following table shows the largest canyons in the world.

Comets

From time to time, **comets** streak through the sky, close enough for you to see them with the naked eye. A

GOD MADE THE WORLD

Largest Canyons in the World

Canyon	Location	Measurements	Runs Into
Yarlung Tsangopo	Tibet, Asia	Depth 6,009 m, Length 505 km	Indian Ocean
Capertee Valley	Australia	Width 135 km	Pacific Ocean
Kali Gandaki Gorge	Nepal, Asia	Depth 6,800 m, Length 627 km	Indian Ocean
Grand Canyon	US, N. America	Depth 1,828 m, Length 445 km	Pacific Ocean
Fish River Canyon	Namibia, Africa	Depth 550 m, Length 160 km	Atlantic Ocean

Grand Canyon, Arizona, USA

CHAPTER 3: GOD MADE THE EARTH

comet is a large chunk of ice, dust, and rock particles—a sort of dirty snowball traveling through our solar system. Most of the time, these comets revolve around the sun, just like our earth. So far, scientists have counted 6,600 comets in our solar system. The largest are 10-20 miles across, about the size of a small island in the middle of the ocean.

As some comets come closer to the sun, a tail becomes visible to those watching from the earth. Parts of the comet (including the icy parts) vaporize into dust and gases as it travels along.

Where did the mysterious water in comets come from? Unbelieving scientists are hoping that there is life on other planets. Since life is dependent upon water, these men suggest the water came from some other planet. They imagine that life somehow just appears all by itself on other planets. They do not believe in the Creator who made life and gave us water on this earth.

The best explanation for the mysterious ice in comets goes back to the worldwide flood. When the earth's crust broke apart, and the waters in the deep exploded into the air, some of all that dirt, magma, and water shot up into space. Immediately,

Comet

the water turned into ice and comets formed. It is also possible that some of the chunks of the debris hit the moon and Mars. That would provide something of an appearance of water in spots on these celestial bodies.

Unbelieving scientists are also baffled over the moon's surface. The near side of the moon facing the earth has larger pockmarks or bigger basins than the other side facing away from us. There are eight big basins, larger than 186 miles (300 km) on the side that faces us. There's only one big basin on the other side. Where would these large pieces of debris that hit the moon come from? Since this is the side that faces the earth, it makes sense that the debris came from the earth during the worldwide flood. Huge chunks of earth's

87

GOD MADE THE WORLD

crust and mantle could very well have flung out into space and hit the moon causing these big basins. This is the best way to explain what happened to the moon. Our world was very much impacted by that big flood God sent over 4,000 years ago. But unbelieving scientists do not accept this major catastrophe which we read about in Genesis 7.

Asteroids and Meteorites

God is our refuge and strength,
A very present help in trouble.
Therefore we will not fear,
Even though the earth be removed,
And though the mountains be carried into the midst of the sea;
Though its waters roar and be troubled,
Though the mountains shake with its swelling. Selah.
There is a river whose streams shall make glad the city of God,
The holy place of the tabernacle of the Most High.
God is in the midst of her, she shall not be moved;
God shall help her, just at the break of dawn.
(Psalm 46:1-5)

An **asteroid** is a large body of rock and metal which travels around the sun, sized from 1/2 mile to 590 miles in length. There are over eight hundred of these asteroids that fly near the earth.

A **meteoroid** is a smaller chunk of rock that breaks off of an asteroid or a comet (up to 3 feet in diameter). If it enters the earth's atmosphere, and burns up it is called a **meteor**. Believe it or not, a total of 48 tons of meteoroids fly into the earth's atmosphere every day. That's a lot of cosmic rocky junk! Most of it burns up, and you may see their

Asteroid

CHAPTER 3: GOD MADE THE EARTH

streaks through the night sky. Sometimes these rocks will hit the ground, and that's what we call **meteorites**.

The largest meteorite ever found is called "The Hoba." It was a 66-ton piece of rock found in a farmer's field in Namibia, Africa in 1920. It was made up of the typical elements we find on the earth, elements God created at the beginning—84% iron, 16% nickel, along with cobalt and a few other metals.

Some of the older and larger craters on Earth may have been formed during the worldwide flood when large chunks of earth blasted out and then crashed back down. The largest crater in the world is in the Vredefort Crater in South Africa.

It is 185 miles across by some estimates.

Some people may wonder if God might accidentally allow an asteroid to careen into the earth, and destroy everything. Unbelieving scientists want to look to organizations like NASA in order to protect the world from this destruction. But, we are not worried about this. Of course, nothing happens accidentally when God is in control of all things. He would never allow an asteroid to destroy the world. God loves His people. His Son, the Lord Jesus Christ died for His church, and He has promised to be with us to the end. We don't have to be afraid even if the mountains were thrown into the oceans. God is right there with the church, and this church will not be moved, no matter what happens.

Astronomers keep an eye on asteroids heading towards the earth. The next big one coming near us in AD 2028 won't even come close to the moon. This asteroid (1997 xf11) is about one mile wide, and it's traveling at 30,000 mph. If this big rock hit New York City, it would destroy everything

Meteorite

GOD MADE THE WORLD

from Washington DC to Boston—a population of about 50,000,000 people. An asteroid sixty miles in length would destroy the whole earth.

Another big asteroid is scheduled to come close to the earth on April 13th, 2029. Asteroid Apophis is 1,213 feet wide (370 m). That's about the size of four football (or soccer) fields. Astronomers estimate it will come within about 18,000 miles (31,000 km) from the earth. That's pretty close—only one-tenth the way to the moon. If it were to hit the earth, it would produce an explosion about like ten of our most powerful nuclear bombs.

Before you get too worried though, remember that not one person has been killed by a meteorite as far as we know, in a thousand years. At this point, astronomers tell us they don't see any large objects coming close to the earth for several hundred years.

It is not for us to fear these meteoroids and asteroids. We must not fear floods and other natural disasters. However, we should fear God. We should be aware that He will destroy the world at the end, and bring every person to His judgment. The epistle of 2 Peter tells us of this coming destruction on the whole world:

> Knowing this first: that scoffers will come in the last days, walking according to their own lusts, and saying, "Where is the promise of His coming? For since the fathers fell asleep, all things continue as they were from the beginning of creation." For this they willfully

Meteor Crater, Arizona, USA

CHAPTER 3: GOD MADE THE EARTH

forget: that by the word of God the heavens were of old, and the earth standing out of water and in the water, by which the world that then existed perished, being flooded with water. But the heavens and the earth which are now preserved by the same word, are reserved for fire until the day of judgment and perdition of ungodly men.... But the day of the LORD will come as a thief in the night, in which the heavens will pass away with a great noise, and the elements will melt with fervent heat; both the earth and the works that are in it will be burned up. Therefore, since all these things will be dissolved, what manner of persons ought you to be in holy conduct and godliness. (2 Peter 3:3-7,10-11)

Fossils

Now the flood was on the earth forty days... All in whose nostrils was the breath of the spirit of life, all that was on the dry land, died. So He destroyed all living things which were on the face of the ground: both man and cattle, creeping thing and bird of the air. They were destroyed from the earth. Only Noah and those who were with him in the ark remained alive. (Genesis 7:17,22-23)

True to the word of God, the flood killed billions of animals very quickly. When these animals were captured in the muddy waters, many of them were fossilized. This means that they were first encased in mud. Then, the heat and pressure on the organisms changed them into a hardened, carbonized form. This left an impression on the sedimentary rock that is preserved for us to look at thousands of years later. These are called **fossils**. There are a few other ways that fossils are preserved, such as petrifaction and casting. You can find fossils all over the world, because the flood waters covered the whole earth. Researchers have even found shellfish at the top of Mount Everest, the tallest mountain on Earth. Of course, these shellfish did not crawl all

Dinosaur Fossil in Rock

GOD MADE THE WORLD

the way up there to die. They were carried by the flood waters in layers of rock which were deposited on the mountain about 4,000 years ago.

Unbelieving scientists tell us that there were many floods that carbonized these ocean creatures and other animals over millions of years. The better explanation is that one gigantic flood did the work of fossilizing so many billions of animals, all at one time. Localized, smaller floods can also fossilize animals here and there around the world.

God Gave Fuel, Coal, and Oil

Here is what I have seen: It is good and fitting for one to eat and drink, and to enjoy the good of all his labor in which he toils under the sun all the days of his life which God gives him; for it is his heritage. As for every man to whom God has given riches and wealth, and given him power to eat of it, to receive his heritage and rejoice in his labor—this is the gift of God. (Ecclesiastes 5:18-19)

By our Creator God's amazing mercies and generous consideration for humans, He gave us sources

Oil Pump Jack in Alberta, Canada

CHAPTER 3: GOD MADE THE EARTH

of energy on the earth to last for a long time. He gave us trees, which we cut up into logs and burn in fireplaces. But, He also gave us coal and oil. Both of these can provide heat and energy for human use.

Oil is a little bit of a mystery. Nobody really knows for sure how it was made. The safest thing to say is that God put it under the ground for us to use. For a long time, scientists thought that oil and gas were created by dead organic matter from the earth. However, fewer scientists hold to that theory now. God might have created them at the beginning, just like He created copper, iron, gold, and silver. During the flood, the oil and gas were probably brought closer to the crust of the earth, so that we could access these energy sources.

Actually, there are seven common types of oil and gas. These are called **hydrocarbons**, because they are made up of two basic substances which God created—hydrogen and carbon.

Usually, geologists find reserves of oil and gas about two to three miles under the surface of the earth. The gas and oil just sit there in cracks and crevices of porous rock. There may be some water there too, but the gas and oil float on water and so they will sit on top, until oil companies start pumping them out.

Oil/Glass	Form
Methane	Natural gas
Ethane	Natural gas
Propane	Gas, although it can become liquid if you put it under pressure
Butane	Gas, although it can become liquid if you put it under pressure
Pentane	Liquid
Hexane	Liquid
Octane	Liquid

Since 1920, oil and gas have been essential for heating, for air travel, for cars and trucks, and most of our energy needs. This resource has been a great blessing from God for modern life all over the world. The following chart shows how much of the world is dependent on oil, gas, and coal. As of 2010, about 87% of the world's energy comes from these fuels.[4]

Countries with the Highest Coal and Oil Reserves

Coal	Percentage of World Reserves[5]	Oil	Billions of Barrels in Reserve[6]
United States	24%	United States	260
Russia	15%	Russia	250
Australia	14%	Saudi Arabia	210
China	13%	Canada	160
India	9%	Iran	140
Indonesia	3.5%	Brazil	120
Germany	3.4%	Iraq	115
Ukraine	3.3%	Venezuela	90

When Will We Run Out of Coal and Oil?

The official numbers for oil reserves in the earth, as of 2019, is about 1.7 trillion barrels. That should last us until AD 2069. However, there may be more reserves of oil and gas to find in the years to come. We'll see. On the other hand, there are about 1.1 trillion tons of coal left in reserves—that's enough to last about 150 years, well after your lifetime.

The countries with the highest reserves of coal and oil, are listed in the table above. It is interesting that God blessed the United States with the highest reserves of all the nations in the world.

Searching for More Fuels and Energy Sources

If the Lord Jesus Christ does not return before the earth runs out of oil, then

CHAPTER 3: GOD MADE THE EARTH

the world will have to find other kinds of energy to fuel cars and heat homes. This will be a challenge for young people who are just now learning about science.

Since you cannot put coal in your car and it is hard to power a car with a windmill sitting on top of it, car companies have developed electric cars. These cars rely on electricity that comes from large coal-burning plants. Unless scientists can come up with another form of gas or combustible stuff that can burn in a car engine, most cars will have to run on electricity in the future. Here are some of the ideas scientists are working with, as they try to find ways to use God's resources to provide the world with energy.

1. Scientists are looking into using silicon found in sand as a fuel. There is a lot of sand in the world, and this may be a good way to develop a fuel. No one has yet figured out a way to turn sand into fuel without spending a ton of money doing it.

2. Ethanol made out of sugar cane or corn is mixed into gasoline. This fuel is already being used all over the world. It takes a lot of work to grow

Trans-Alaska Pipeline

these crops, so this is not a very good way to create fuel.

3. Geothermal energy takes heat the from the ground (the earth's crust and mantle), and uses it to heat homes, and make electricity. This is probably the cheapest and most reliable of the alternative kinds of energy.

4. Nuclear energy is already used for 6% of the world's needs. Scientists have developed smaller nuclear plants that can power a little town. These may be a safer approach to nuclear energy. Also, scientists are seeing breakthroughs in developing nuclear fusion, a better form of energy than the fission plants currently used.

5. Solar, wind, hydro-electric (river dams), and tide-generated power are used here and there around the world. These forms are still expensive, but scientists are still experimenting with them, and finding better ways to produce them. These creative forms of harnessing energy have come down in cost over the last few decades. You can see that the cheapest energy is still nuclear power and coal power.

Form	Production Cost as of 2018
Nuclear	.15 / kW-hr[7]
Coal	.10 / kW-hr
Natural Gas	.06 / kW-hr
Hydroelectric	.05 / kW-hr[8]
Bioenergy	.06 / kW-hr
Onshore Wind	.04 / kW-hr
Geothermal	.09 / kW-hr
Photovoltaic Solar	.04 / kW-hr

How Much Would Your Family Pay for This Electricity?

The average home uses about 900 kW-hours of electricity each month. So that means your electric bill would be $18 a month with nuclear energy. If energy companies relied on solar, your electric bill would be $81 a month. So you can see that there is still a big difference in costs between the different types of energy.

Tihange Nuclear Power Station in Huy, Belgium

Praise God for His Resources

The craftsman stretches out his rule,
He marks one out with chalk;
He fashions it with a plane,
He marks it out with the compass,
And makes it like the figure of a man,
According to the beauty of a man, that it may remain in the house.
He cuts down cedars for himself,
And takes the cypress and the oak;
He secures it for himself among the trees of the forest.
He plants a pine, and the rain nourishes it.
Then it shall be for a man to burn,
For he will take some of it and warm himself;
Yes, he kindles it and bakes bread;
Indeed he makes a god and worships it;
He makes it a carved image, and falls down to it. . .
"Remember these, O Jacob,
And Israel, for you are My servant;
I have formed you, you are My servant;
O Israel, you will not be forgotten by Me!
I have blotted out, like a thick cloud, your transgressions,

GOD MADE THE WORLD

And like a cloud, your sins.
Return to Me, for I have redeemed you."
Sing, O heavens, for the LORD has done it!
Shout, you lower parts of the earth;
Break forth into singing, you mountains,
O forest, and every tree in it!
For the LORD has redeemed Jacob,
And glorified Himself in Israel.
(Isaiah 44:13-15,21-23)

God has given us wood, coal, oil, and metals for our use. Yet, man often turns these good gifts into gods. Men will use one of God's trees to build a chair, and a warming fire. Then, he makes an idol out of the rest of the tree. What foolishness! Today, men value their cars and houses more than they worship the Creator.

This passage tells us that every tree of the forest sings praises to God the Creator. Let us sing praises to God too. Let us thank Him for His good gifts!

Pray
- Let us thank the Lord for the blessings of energy, electricity, and heat!
- Thank Him for burying all that coal and gas under the ground for your family, and for the rest of the world.
- Praise God for His power and goodness to protect our earth from flying asteroids! Praise Him for the mighty mountains, and thank Him for His strong protection over us!
- Acknowledge His great judgments in the earth, that come with the worldwide flood and with earthquakes and volcanoes. Then, thank Him for saving Noah and his family from the flood. Thank Him for saving us from destruction by sending His Son to die on the cross.

Do

Hurricane and Volcano Preparedness

First, you should be aware of the earthquake fault lines and the active volcanoes in your area. Be aware of flood plains. Where is flooding most likely to occur?
1. Research the earthquake fault lines in your area. Where are the nearest fault lines to where you live? What are the more active volcanoes in your state or nation?
2. Identify the places in your home where you would go if there was an earthquake in your area. What would you stay away from, in such a case?
3. Ask your parents to show you the location of the gas and electric shut-offs for your home.

Alternative Energy
1. Build your own windmill using a simple kit available online.
2. Create your own solar power using a simple kit available online.

Climb a Mountain

Go hike a hill or a mountain with a parent, and give God the glory for His power and beauty!

Watch

To watch the recommended videos for this chapter, go to generations.org/GodMadeTheWorld and scroll down until you find the video links for Chapter 3. Our editors have been careful to avoid films with references to evolution. However, we would still encourage parents or teachers to provide oversight for all internet usage. These videos may not give God the glory for His amazing creative work, so the student and parent/teacher should respond to these insights with prayer and praise.

Yosemite Valley, California, USA

Chapter 4
GOD MADE MATTER

> For the wrath of God is revealed from heaven against all ungodliness and unrighteousness of men, who suppress the truth in unrighteousness, because what may be known of God is manifest in them, for God has shown it to them. For since the creation of the world His invisible attributes are clearly seen, being understood by the things that are made, even His eternal power and Godhead, so that they are without excuse... (Romans 1:18-20)

We can learn about God by studying everything He has created. We can see His power and His wisdom in this amazing created universe. In fact, Romans 1 tells us that it is wrong to ignore the Creator as we look at the creation. It is to be rebellious against God. The creation shouts His wonderful character! Let us learn about God as we study His world.

Of course, you cannot see God because He is a Spirit. God was at the beginning and He has always been there. Nothing made God, because there was never a time when God was not there. God made everything except for Himself. He made everything that you can see and everything you cannot see. We usually call this the material and the immaterial, or the visible and the invisible.

The invisible parts of God's creation are things like angels and the human

GOD MADE THE WORLD

soul. You are made up of more than just a material body. You are more than fingers and toes, and skin and bones. You were created with an invisible soul that makes up your personality.

There are two kinds of materials in the world—the animate and the inanimate. The animate is that which is living, and the inanimate is that which is not living. Here we are mainly looking at the inanimate creation of God.

The technical word for any material thing is **matter.** The most basic way to measure matter is by its **mass.** A little ping pong ball has less mass than a bowling ball. There is more "stuff" in a bowling ball so we say that it has a "higher mass."

Tiny Building Blocks

The LORD by wisdom founded the earth; By understanding He established the heavens. (Proverbs 3:19)

All of the material world is made up of tiny building blocks. Just as children will build cars and houses out of Lego® bricks, all creation is made up of these tiny building blocks called **atoms**.

The atom is so small, you could fit 200 trillion of them into the period at the end of this sentence.

An atom is made up of other smaller pieces of stuff (or matter). But if you were to pull one of those pieces off of the atom it could not be used to build anything. So the atom is the basic building block of all things. The atom is made up of the following:
- Protons
- Neutrons
- Electrons

The main difference between these pieces of matter is their electrical

Bowling Ball and Ping Pong Ball

CHAPTER 4: GOD MADE MATTER

charge. The proton has a positive charge. The neutron has a neutral charge (no charge), and the electron has a negative charge.

The protons and neutrons make up the core or the nucleus of the atom, and the electrons fly super fast around the core (sort of like the way the earth revolves around the sun).

Most of the time, atoms hold the same number of negatively charged electrons and positively-charged protons. However, **ionized atoms** have unequal numbers of electrons and protons. We're all familiar with ionized material in the form of **static cling.** When you rub a balloon in your hair, atoms become ionized and you create this electro-static condition.

Negatively-charged electrons repel each other, but they stay close to the positively-charged nucleus because negative charges are attracted to positive charges. Opposites attract. This attraction between the electrons and the nucleus of the atom is called the **electromagnetic force.** It is this force that provides the centripetal movement of the electrons in orbit around the nucleus.

These parts of the atom (protons, neutrons, and electrons) are super small. An electron is about 0.0000000000001

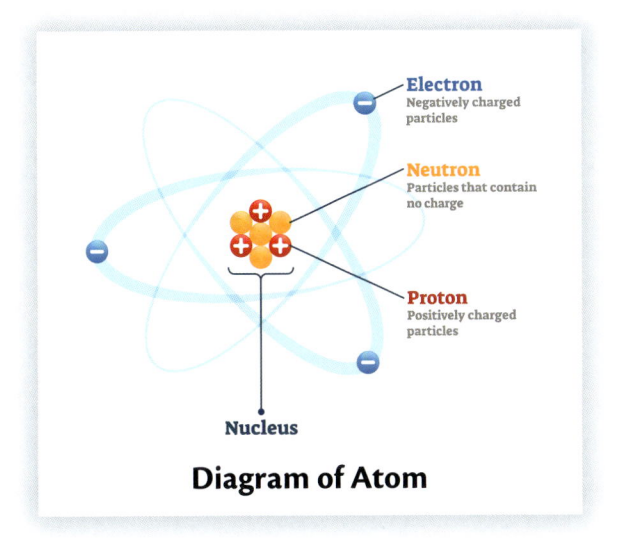

Diagram of Atom

- Electron: Negatively charged particles
- Neutron: Particles that contain no charge
- Proton: Positively charged particles
- Nucleus

centimeters across, or 2,000 billion times smaller than one grain of sand. If an atom were the size of a football field, the electron would be about the size of a football. Using a **quantum microscope**, scientists have been able to take pictures of an atom.

If you were to take a grain of sand out of the sandbox and cut it apart, what would you find? What if you kept cutting it into smaller pieces. What is this little piece of sand made of, and how does it stick together like this?

That grain is made of several minerals, but mostly of a material called **quartz**. If you were to break the grain of sand apart, you would find **molecules** made up of silicon and oxygen atoms. Molecules are

GOD MADE THE WORLD

combinations of atoms. They stick together when the atoms share electrons in what is known as a **chemical bond.**

Scientists have studied protons and neutrons, and discovered that they are made up of super tiny objects called **quarks.** You will never find a quark sitting all by itself. You cannot separate it from other quarks. But there are six different kinds of quarks. They are named Up, Down, Strange, Charm, Bottom, and Top. So far, scientists have figured out that it takes two up quarks and one down quark to make a proton, and two down quarks and one up quark to make a neutron.

You may remember that the Lord Jesus created four forces which He uses to hold the whole world together. The force of gravity keeps our earth in place, revolving around the sun. Gravity keeps things resting on the earth, so things don't float off into space. The force that keeps the atom together is called the **electromagnetic force.**

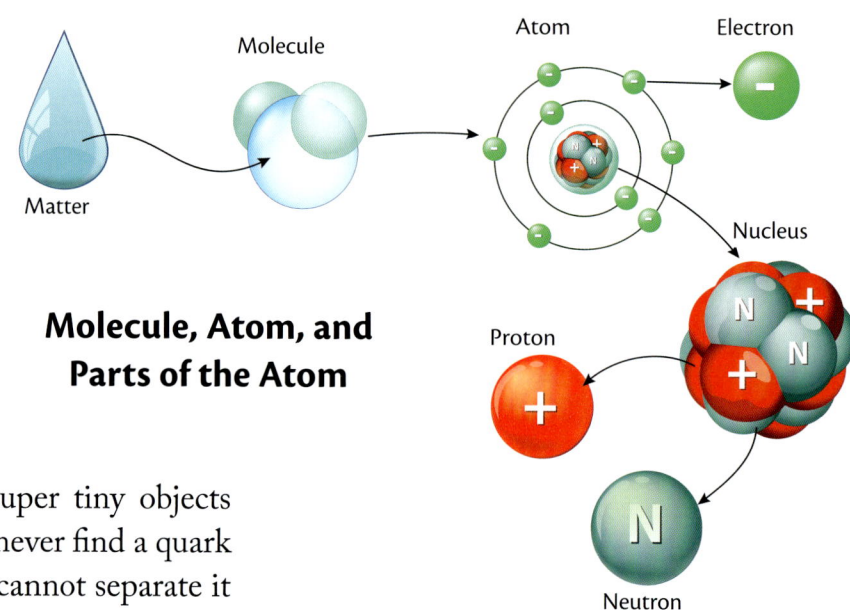

But, what mysterious forces hold the atom's nucleus together? There is the **Strong Force** and the **Weak Force.**

The Weak Force is stronger than gravity, but it can only work at very short distances between different things. Somehow, God uses the Weak Force to change a proton into a neutron or an electron. This starts a decaying process which will change the material of the atom into something different.

The Strong Force is the strongest force in the universe. God made it to hold the nucleus together so the neutrons and

Molecule, Atom, and Parts of the Atom

104

CHAPTER 4: GOD MADE MATTER

protons don't drift apart. How would you hold a positively charged proton against a neutral neutron? Remember that positive charges will attract negative charges. But, positive charges are not attracted to neutral charges like what you find in a neutron. This reveals the amazing wisdom of our Creator God. These positive particles are held together in a very tight space and yet they will not repel each other.

The Strong Force (also called the **Nuclear Force**) is what holds the quarks together. This force is strong enough to hold together the protons in the nucleus that want to repel each other. The force of gravity decreases when you separate two objects from each other. However, God designed the Strong Force very differently. The farther the quarks separate in a proton or neutron, the stronger the force becomes. What does this mean? Of course, God wants to keep His creation sticking together. We don't want our bodies, the trees, rocks, and cars falling apart or disintegrating all the time.

Praise God for these amazing forces He has created in this fantastic world of ours! These forces are so mysterious; the brightest scientists in the world cannot figure out what produces these forces.

> Oh, the depth of the riches both of the wisdom and knowledge of God! How unsearchable are His judgments and His ways past finding out! (Romans 11:33)

How Does the Lord Hold Everything Together?

> For it was fitting for Him [Jesus], for whom are all things and by whom are all things, in bringing many sons to glory, to make the captain of their salvation perfect through sufferings. For both He who sanctifies and those who are being sanctified are all of one, for which reason He is not ashamed to call them brethren. (Hebrews 2:10-11)

Don't forget, that it is in the Lord Jesus that "all things consist" (Colossians 1:17). At the beginning, He created all things and by Him all things continue to exist. So how does our Lord hold a piece of sand together? How does He hold the whole world together? Let us review:

All of the sand, the dirt, the rocks, and the water (as well as ourselves) do not float away from the earth because of gravity. **Gravitational force** keeps the

Without Gravity, Everything Floats Away

earth together.

The molecule is held together by a chemical bond, a sharing of electrons between atoms.

The atom is held together by **electromagnetic force** between negative electrons and the positive nucleus.

The nucleus is held together by the **Strong Force** acting between the quarks.

And that is how Jesus holds matter in this world together!

The Smallest Building Blocks for the Whole Universe

So now, the whole universe is made out of these things. The stars, the sun, the moon, the earth, the sand, the water, the plants, the animals, and you are made up of these things. If you were to break the grain of sand down into smaller and smaller pieces, here is what you

CHAPTER 4: GOD MADE MATTER

would find:
- Grains of sand
- Molecules made of silicon and oxygen atoms
- Electrons and the nuclei of the atoms
- Protons and neutrons
- Quarks

The Elements

The fear of the LORD is clean, enduring forever;
The judgments of the LORD are true and righteous altogether.
More to be desired are they than gold,
Yea, than much fine gold;
Sweeter also than honey and the honeycomb.
Moreover by them Your servant is warned,
And in keeping them there is great reward.
(Psalm 19:9-11)

There are 41 different solid colors of Lego® bricks. When children build stuff out of Legos® they will use all these different bricks to make little cars, houses, and other creations. Now, as far as we know, God made 118 different kinds of atoms in the world. Everything in the whole universe is built out of these 118 different kinds of building blocks. There is much variety in the world. Look around you. Is everything the same color and the same shape? There are soft things and hard things. There are green things and brown things. There are trees, rocks, sand, frying pans, human beings, dogs, and cats. All of these things are

107

made out of different kinds of materials. Your body is not made out of the same thing that frying pans are made of. God uses different materials to build all these different creations. These various kinds of materials or atoms are called **elements**.

The elements are presented on a **periodic table** contained on the next page. The number at the top indicates the number of electrons in the atom. Each element has a symbol by which we can identify what it is.

The most common elements in the universe are hydrogen, helium, oxygen, and silicon.

Families of Elements

God has created different families of elements, all very useful for mankind. There is a family of elements called **halogens.** They are highly reactive, but they become very stable once they combine with another element. The halogens have only seven electrons floating around their outer shell, and they really would like an eighth. These atoms are not happy being by themselves, and they are not seen in nature by themselves. This group of elements include fluorine (atomic symbol F), chlorine (Cl), bromine (Br), iodine (I), and astatine (At).

Chlorine combines with sodium to make salt.

Iodine combines with potassium to make another kind of salt (KI).

People use chlorides and iodides for diet and health reasons.

Some compounds use halogens as fire-retardants, refrigerants (freon), and non-stick coating for pots and pans (Teflon).

Most gases are colorless, but fluorine gas is pale yellow, and chlorine gas is a yellowish green. Iodine vapors are colored deep purple.

While halogens are unstable, the **noble gases** are highly stable. They have a perfect set of eight electrons floating around in their outer shell, and they are happy to stay that way. Helium (He), neon (Ne), and argon (Ar) are examples of nobles gases.

The **alkali metals** on the far left of the Periodic Chart include lithium, sodium, and potassium. They are super soft metals, shiny and lustrous. The **alkaline earth metals** are more dense, heavier, and harder than the alkalis. These include calcium (Ca) and magnesium (Mg).

God created different materials for different uses. What a blessing that we can use Teflon on cooking pans! This compound made of fluorine (F) and carbon (C) makes it so much easier to clean the remains of scrambled eggs remains off the pans. Teflon was "accidentally" discovered in a research laboratory by a 27-year-old chemist named Roy J. Plunkett. He was raised on a farm by Christian parents, and discovered Teflon on April 6, 1938 while working on a non-poisonous refrigerant for the DuPont company. Often, God reveals exciting discoveries to scientists who are exploring His world. "It is the glory of God to conceal a matter, but the glory of kings is to search out a matter" (Proverbs 25:2).

Molecules and Compounds

Stand up and bless the LORD your God
Forever and ever!
Blessed be Your glorious name,
Which is exalted above all blessing and praise!
You alone are the LORD;
You have made heaven,
The heaven of heavens, with all their host,
The earth and everything on it,
The seas and all that is in them,
And You preserve them all.
The host of heaven worships You.
(Nehemiah 9:5-6)

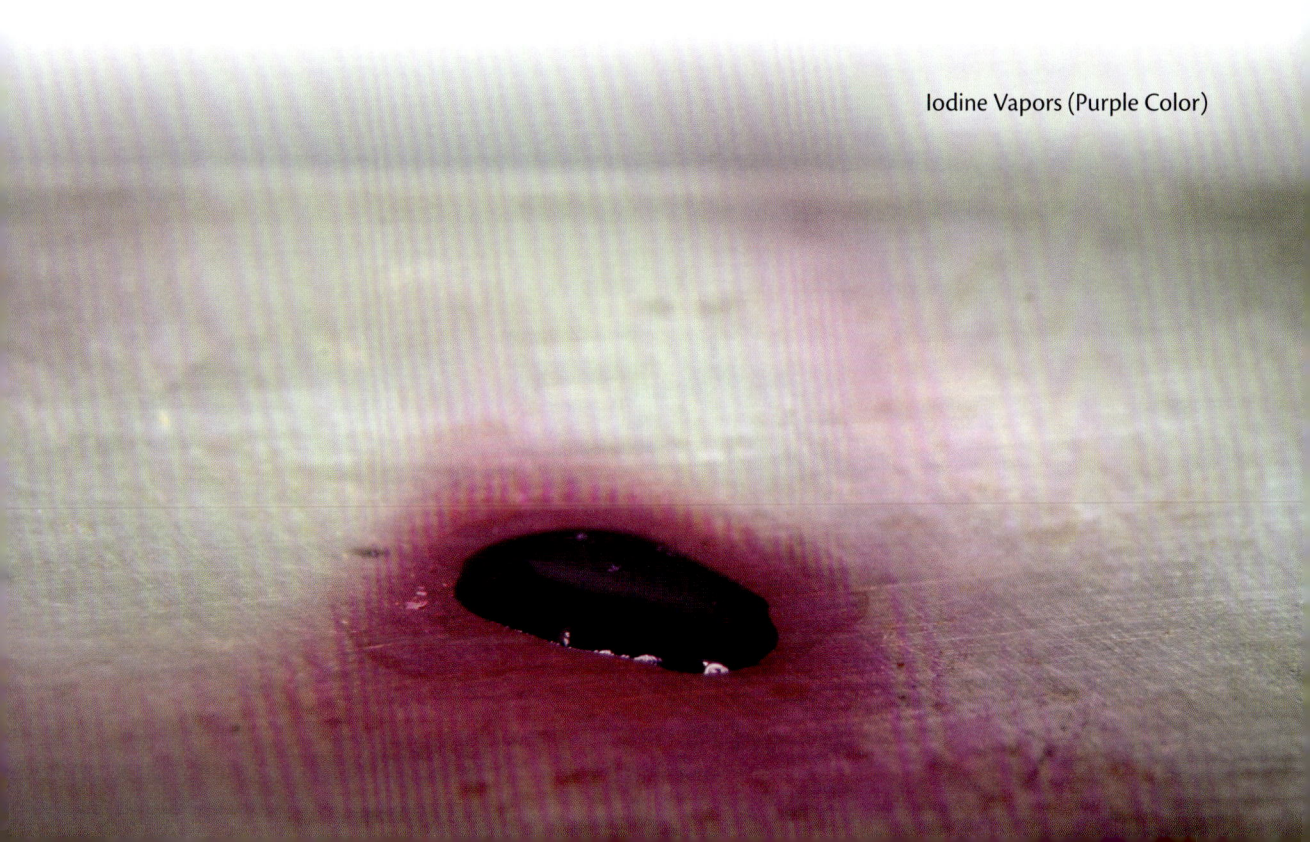

Iodine Vapors (Purple Color)

GOD MADE THE WORLD

Periodic Table of Elements

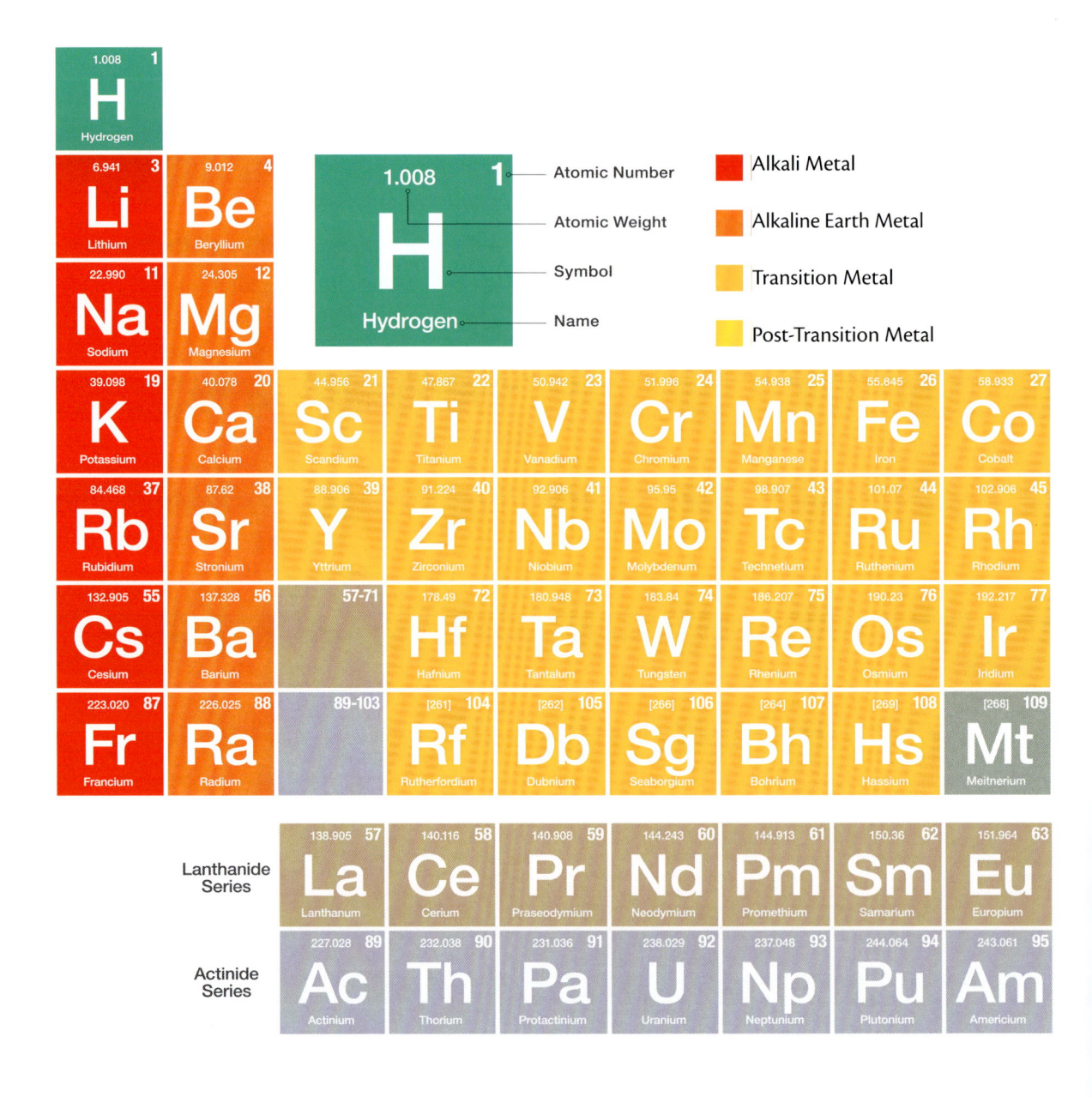

1.008 **1** **H** Hydrogen								
6.941 **3** **Li** Lithium	9.012 **4** **Be** Beryllium		1.008 **1** **H** Hydrogen — Atomic Number / Atomic Weight / Symbol / Name			Alkali Metal / Alkaline Earth Metal / Transition Metal / Post-Transition Metal		
22.990 **11** **Na** Sodium	24.305 **12** **Mg** Magnesium							
39.098 **19** **K** Potassium	40.078 **20** **Ca** Calcium	44.956 **21** **Sc** Scandium	47.867 **22** **Ti** Titanium	50.942 **23** **V** Vanadium	51.996 **24** **Cr** Chromium	54.938 **25** **Mn** Manganese	55.845 **26** **Fe** Iron	58.933 **27** **Co** Cobalt
84.468 **37** **Rb** Rubidium	87.62 **38** **Sr** Strontium	88.906 **39** **Y** Yttrium	91.224 **40** **Zr** Zirconium	92.906 **41** **Nb** Niobium	95.95 **42** **Mo** Molybdenum	98.907 **43** **Tc** Technetium	101.07 **44** **Ru** Ruthenium	102.906 **45** **Rh** Rhodium
132.905 **55** **Cs** Cesium	137.328 **56** **Ba** Barium	57-71	178.49 **72** **Hf** Hafnium	180.948 **73** **Ta** Tantalum	183.84 **74** **W** Tungsten	186.207 **75** **Re** Rhenium	190.23 **76** **Os** Osmium	192.217 **77** **Ir** Iridium
223.020 **87** **Fr** Francium	226.025 **88** **Ra** Radium	89-103	[261] **104** **Rf** Rutherfordium	[262] **105** **Db** Dubnium	[266] **106** **Sg** Seaborgium	[264] **107** **Bh** Bohrium	[269] **108** **Hs** Hassium	[268] **109** **Mt** Meitnerium

Lanthanide Series: 138.905 **57** **La** Lanthanum | 140.116 **58** **Ce** Cerium | 140.908 **59** **Pr** Praseodymium | 144.243 **60** **Nd** Neodymium | 144.913 **61** **Pm** Promethium | 150.36 **62** **Sm** Samarium | 151.964 **63** **Eu** Europium

Actinide Series: 227.028 **89** **Ac** Actinium | 232.038 **90** **Th** Thorium | 231.036 **91** **Pa** Protactinium | 238.029 **92** **U** Uranium | 237.048 **93** **Np** Neptunium | 244.064 **94** **Pu** Plutonium | 243.061 **95** **Am** Americium

CHAPTER 4: GOD MADE MATTER

- Metalloid
- Polyatomic Nonmetal
- Diatomic Nonmetal
- Noble Gas
- Lanthanide
- Actinide
- Unknown Properties

						4.003 2 **He** Helium
10.811 5 **B** Boron	12.011 6 **C** Carbon	14.007 7 **N** Nitrogen	15.999 8 **O** Oxygen	18.998 9 **F** Fluorine	20.180 10 **Ne** Neon
26.982 13 **Al** Aluminum	28.086 14 **Si** Silicon	30.974 15 **P** Phosphorus	32.066 16 **S** Sulfur	35.453 17 **Cl** Chlorine	39.948 18 **Ar** Argon

58.693 28 **Ni** Nickel	63.546 29 **Cu** Copper	65.38 30 **Zn** Zinc	69.723 31 **Ga** Gallium	72.631 32 **Ge** Germanium	74.922 33 **As** Arsenic	78.971 34 **Se** Selenium	79.904 35 **Br** Bromine	84.798 36 **Kr** Krypton
106.42 46 **Pd** Palladium	107.868 47 **Ag** Silver	112.411 48 **Cd** Cadmium	114.818 49 **In** Indium	118.711 50 **Sn** Tin	121.760 51 **Sb** Antimony	127.6 52 **Te** Tellurium	126.904 53 **I** Iodine	131.294 54 **Xe** Zenon
195.085 78 **Pt** Platinum	196.967 79 **Au** Gold	200.592 80 **Hg** Mercury	204.383 81 **Ti** Thallium	207.2 82 **Pb** Lead	208.980 83 **Bi** Bismuth	[208.982] 84 **Po** Polonium	209.987 85 **At** Astatine	222.018 86 **Rn** Radon
[269] 110 **Ds** Darmstadtium	[272] 111 **Rg** Roentgenium	[277] 112 **Cn** Copernicium	Unknown 113 **Uut** Ununtrium	[289] 114 **Fl** Flerovium	Unknown 115 **Uup** Ununpentium	[298] 116 **Lv** Livermorium	Unknown 117 **Uus** Ununseptium	Unknown 118 **Uuo** Ununoctium

| 157.25 64 **Gd** Gadolinium | 158.925 65 **Tb** Terbium | 162.500 66 **Dy** Dysprosium | 164.930 67 **Ho** Holmium | 167.259 68 **Er** Erbium | 168.934 69 **Tm** Thulium | 173.055 70 **Yb** Ytterbium | 174.967 71 **Lu** Lutetium |
| 247.070 96 **Cm** Curium | 247.070 97 **Bk** Berkelium | 251.080 98 **Cf** Californium | [254] 99 **Es** Einsteinium | 257.095 100 **Fm** Fermium | 258.1 101 **Md** Mendelevium | 259.101 102 **No** Nobelium | [262] 103 **Lr** Lawrencium |

111

Normally, you don't get these atoms all by themselves. They usually combine with other elements to form molecules. For example, it takes two oxygen atoms to make the oxygen we find all around us. This is given the formula O_2. Oxygen atoms don't like to be by themselves. Sometimes three oxygen atoms will come together to form a different molecule called **ozone**. The formula is O_3.

When different elements like hydrogen and oxygen bond, the combination is called a **compound**. These compounds are represented by formulas like H_2O for water, or $NaCl$ for salt. This helps us to see what atoms are used to make up the molecule of the compound. For example, a molecule of water is made of two hydrogen atoms and one oxygen atom. A molecule of salt is made of one sodium (NA) atom and one Chloride atom (Cl). Based on a portion of the periodic chart above, see if you can identify the elements used for the following:

$NaHCO_3$—Baking Soda

$C_{12}H_{22}O_{11}$—Cane Sugar

Some elements are more rare and very hard to find. For example, silver is easier to find than gold. That is why gold is about forty times more expensive than silver. If silver is sold for $18 for a one-ounce coin, then gold would sell for about $1,600 for a one-ounce coin.

Astatine is the rarest element in the world. Scientists estimate there are only 30 grams of it in the whole world (at one time). While it can be made artificially, astatine breaks down quickly into another element. If you started with 30 grams, you will only have 15 grams after eight hours, 7.5 grams after another 8 hours, and 3.75 grams after 24 hours.

Yet, more precious than all of these rare substances is the Word of God. The wisdom of God is more profitable to us than all of the gold, silver, and astatine in the world (Proverbs 3:14).

Solids, Liquids, and Gases, and Their Uses

Matter can exist in three unique states—as solids, liquids, and gases. You

Pan with Teflon Coating

Baking Soda

can see liquids and solids pretty easily, but you can't see most gases.

Why are some materials hard and solid like wood and rocks, while other materials are liquids or gases? What makes the difference? The state of matter has everything to do with the motion of the electrons around the nucleus of the atom. If the electrons are traveling super slow, the atoms stick together much better and they form a solid. When electrons are revolving faster and faster, the material turns into a more gooey flux (still a solid). The atoms flow past each other easily—runny and unstable. In the liquid state, the atoms stick to each other a little bit, but they also slip and slide past each other. As the electrons start to move super fast, the matter becomes gaseous. At this point, the atoms or molecules don't really stick to each other at all.

The temperature of the material is really important to determine the state of it. Solids can turn into liquids if you heat them up, and then liquids turn into gases as you apply even more heat. The faster the electrons move in the material, the harder it is for the atoms or molecules to stick close to each other.

The most common example of this is water. If you heat snow and ice up to 32 °F (0 °C), it will melt into water. Heat the water to 212 °F (100 °C), and it will turn into steam (or the gaseous state). When water melts or turns into steam, it has not changed its basic nature. The ice, water, and steam is still water. It does not lose any mass in the process. It has only changed its state.

Changing States by Temperature

Element	Melting Point Turns from Solid to Liquid	Boiling Point Turns from Liquid to Gas
Oxygen	-362 °F (-218 °C)	-297 °F (-183 °C)
Alcohol (Ethanol)	-173 °F (-114 °C)	172 °F (78 °C)
Water	32 °F (0 °C)	212 °F (100 °C)
Lead	328 °F (164 °C)	1,750 °F (954 °C)
Iron	2,800 °F (1538 °C)	5,182 °F (2861 °C)

You can't smash wood or water into a smaller and more confined space very well. However, gases can be pressurized or compressed. Also, gases expand quite a bit when they are heated up. If you were to heat up a helium balloon, it would get bigger and bigger. It takes more volume to contain the same amount of helium. If you were to cool it down, the balloon would get smaller and smaller.

It's hard to imagine that invisible gases actually have mass or weight. But suppose you were to weigh a balloon on a very accurate scale before you blew it up. The balloon would weigh more after you filled it with air and tied it off. That's because air has mass and weight. It's still pretty light compared to milk. One hundred gallons of air would weigh about one pound. One hundred gallons of milk would weigh about 215 pounds.

Air

"The Spirit of God has made me,
And the breath of the Almighty gives me life." (Job 33:4)

We breathe air all day long. You can't go very long without breathing. Your body needs air to live. Your brain needs air to think. Your lungs need air to do their job, which is to move oxygen into red blood cells. These red blood cells carry that oxygen into every nook and cranny of the body. This oxygen helps process your food into energy, so you can live. We breathe about 2,900 gallons of air a day (11,000 liters). Imagine drinking that much milk in a day! That's about 85,000,000 gallons in a lifetime. Oh that we would remember it is God who has given us all of that air

The Father of Chemistry

This chapter has introduced the subject of **chemistry**. Who was the most important scientist, who developed this material for the modern world? It was Robert Boyle who figured out that there were elements that could not be broken down into simpler material. He introduced the term "chemical analysis" and defined compounds and mixtures. He is best known for Boyle's Law which relates the volume of a gas to its pressure.

Boyle was a very committed Christian raised by Puritan parents. He declared that the purpose of his study of the world was to give glory to God.

"When, in a word, by the help of anatomical knives, and the light of chymical furnaces, I study the book of nature. . .I find myself oftentimes reduced to exclaim with the Psalmist, 'How manifold are Thy works, O LORD! in wisdom hast Thou made them all!' (Psalm 104:24)"

Robert Boyle wrote that there are

Robert Boyle (1627-1691)

two goals for the study of chemistry and the world of science. First, it is to witness the manifestation of God's glory (Ps. 19:1, Prov. 16:4, Rom. 11:36). And, secondly, it is to subdue creation (Gen. 1:28-29) and use it for the benefit of mankind (Gen. 1:14-16; Is. 45:18). The study of God's creation should not chase people away from God, but rather draw them to Him.

to breathe. Should we use some of it to thank Him for His many gifts?

The longest anybody has ever gone without breathing is 24 minutes and 3.45 seconds. Aleix Segura Vendrell set the record in Barcelona, Spain, on February 28, 2016. Man cannot live for long without breathing God's air.

Air is a **mixture** of different gas molecules. There are varying amounts of water and dust in the air that we breathe. The sixteen different substances listed are in every breath of air you take. However, they do not combine to create one big molecule or compound. They are mixed together like chocolate chips, flour, and sugar are mixed into a bowl to make chocolate chip cookies.

You will notice that there is a nitrogen or oxygen atom, and then there is also a nitrogen or a oxygen molecule.

Chemical Reactions and Mixtures

"You are the salt of the earth; but if the salt loses its flavor, how shall it be seasoned? It is then good for nothing but to be thrown out and trampled underfoot by men." (Matthew 5:13)

Molecule	Symbol	Percentage of the air we breathe
Nitrogen	N_2	78%
Oxygen	O_2	21%
Argon	Ar	0.9%
Carbon Dioxide	CO_2	0.03%
Neon	Ne	0.002%
Methane	CH_4	0.0002%
Helium	He	0.0005%
Krypton	Kr	0.0001%
Hydrogen	H_2	0.00005%
Xenon	Xe	0.000009%
Ozone	O_3	0.000007%
Nitrogen Dioxide	NO_2	0.000002%
Iodine	I_2	0.000001%
Water	H_2O	0 - 5.0%
Dust and Pollution		0.025 - .05%

CHAPTER 4: GOD MADE MATTER

Salt is a wonderful blessing. It enhances flavors in food, making the food taste better. It also preserves things like meat, when there is no way to refrigerate. Christians are a blessing to the world, just like salt is an added blessing to our food. You can make salt by combining sodium and chloride. When you make a compound like salt, the original materials do not hold on to their original properties. By themselves, sodium and chloride are poisonous. Combine these two substances into a salt compound by a chemical process, and the new compound (salt) is not poisonous anymore. Your mom can put it on food without poisoning the family. Do you see that this chemical process works differently than mixing up a salad or blending up a smoothie? If you sprinkled a little poison into the salad or the smoothie, the final product would still be poisonous.

Smoothie Mixture

Have you ever seen an old rusty car? Where does all that orange dust come from? It turns out that oxygen atoms in rain water react with the iron in the steel car. This creates what is officially known as **iron oxide**. If the process continues for

Rusted Truck

GOD MADE THE WORLD

Salt

a long time, you will see the steel turn into these iron oxide flakes. And, you will see holes form in the body of the car.

Baking cookies also changes the chemical nature of the dough. When you heat up a mixture, this can help to bring about the chemical reaction. Have you noticed how baked cookies taste quite a bit different than cookie dough?

So, **mixtures** are not the same thing as substances made by chemical reactions like salt or rust. A mixture is a material made up of two or more different substances which are physically combined.

Blended smoothies are mixtures. Shaving cream is a mixture. Sand is made up of a mixture of a lot of different things, including shells, silica, and little rocks.

Oranges are squeezed of their juice and mixed with water to create orange juice. Sometimes the water separates from the orange juice, and you have to shake it a little bit to mix it up again.

When a chemical reaction happens, you can't separate the parts into the original form anymore. You can't take rust and pull the oxygen and iron back out to make water and steel.

If you mixed a jar full of coins, you could still separate out each kind of coin in piles. You can see that salad dressing mixture of oil and vinegar will separate pretty quickly.

Liquid Solutions, Concentration and Saturation

A liquid solution is a well-mixed mixture. If you mix salt or sugar into water, you can't even see the particles of sugar or salt anymore. It's hard to separate the sugar and salt, unless you boiled the water until it was all gone. Some liquids are really good at dissolving substances like salt and sugar. These liquids are called **solvents**—a liquid that dissolves other substances. Water is one of the very best solvents in the world.

The stuff that is mixed into the solvent is called the **solute.** Think of it this way:

Solvent + Solute = Solution

If you're having a hard time getting a solution to mix up, you can help mixtures along by heating up the solution and stirring it. Put a teaspoon of sugar into your hot tea, and you only need to stir for about 10 seconds to dissolve all of the sugar. Put a teaspoon of sugar into your iced tea, and you have to stir for about 2 minutes to make it all dissolve. Otherwise, the sugar will just sit there at the bottom of the glass.

Gases can also be mixed into liquids. That is how soft drinks are made. Carbon dioxide gas is the most soluble non-toxic gas, and it is easily dissolved into a sweet drink. When the carbon dioxide finds the water molecules, they combine to form carbonic acid, which is safe to drink. These soft drinks have a nice sour taste which complements the sweet taste. As the pop can is opened and poured into a glass, the carbon dioxide bubbles rise to the top. This also produces a tingly sensation on the tongue, which many people find refreshing on a hot day.

The most popular soft drink in the world is still CocaCola®, followed by Pepsi®, Red Bull® (an energy drink), Sprite®, Mountain Dew®, and Dr. Pepper®. In fact, the third largest food and beverage company in all the world is Pepsico® , and the seventh largest is CocaCola®. These carbonated beverages are some of the most popular products in the world, serving billions of people every

Sugar in Iced Tea

GOD MADE THE WORLD

Soda Drinks

day. They are made by mixing a carbon-dioxide and sugar water solution, and some flavorings.

Coffee and tea are also made up of small particles of leaves or beans mixed into water. Besides water, the most popular drink in the US is coffee. Worldwide, the most popular drink is tea—with the world's peoples consuming about 6 billion cups a day. Some people drink strong coffee or tea, while others drink it weak. Usually, scientists refer to this as the **concentration** of the solution (the relative amount of solute in the solution). If you can see the bottom of your glass as you peer into your orange juice, the concentration is very low. There is less solute in the solution. The mixture is diluted. If you cannot see the bottom of your cup as you look into a cup of tea or a glass of orange juice, the concentration is high.

There is a point at which you cannot dissolve any more sugar to water or iced tea. As you keep adding the sugar and stirring, you will see the sugar collecting on the bottom of the glass. This is because you have reached the **saturation point**. The solution is holding all the solute it can handle.

The following lists the most popular solutions people like to drink in the United States.

CHAPTER 4: GOD MADE MATTER

Drink	How much the average American drinks in a year
Coffee	89 gallons
Soft Drinks	49 gallons
Tea	34 gallons
Bottled Water	34 gallons
Beer	20 gallons
Juice	9 gallons
Wine	2.6 gallons

Let us thank God for His provision of interesting and delicious solutions which we may drink. Yet, sometimes men and women are not thankful for God's good gifts. At times, they become addicted to some drinks like coffee or alcohol. When they drink too much alcohol, they actually are sinning against God. This is called the sin of drunkenness. The Bible warns about eating too much food as well, and giving way to gluttony.

Hear, my son, and be wise;
And guide your heart in the way.
Do not mix with winebibbers,
Or with gluttonous eaters of meat;
For the drunkard and the glutton will come to poverty,
And drowsiness will clothe a man with rags.
(Proverbs 23:19-21)

Wine is another solution made up of fruit juices (usually grapes), sugar, yeast, and a few chemicals. The yeast reacts with the sugar, fermenting the sugar into alcohol and carbon dioxide. People were amazed at the miracle Jesus

Coffee Drink Mixture

121

GOD MADE THE WORLD

performed when He made wine at the wedding of Cana. In mere seconds, He turned 150 gallons of water into a very complex, fermented drink mixture that would have taken one to six weeks to accomplish. It would have taken 2,700 pounds of grapes to produce that much wine! This was obviously the work of the Creator Himself, when He walked on earth among us.

On the third day there was a wedding in Cana of Galilee, and the mother of Jesus was there. Now both Jesus and His disciples were invited to the wedding. And when they ran out of wine, the mother of Jesus said to Him, "They have no wine." Jesus said to her, "Woman, what does your concern have to do with Me? My hour has not yet come." His mother said to the servants, "Whatever He says to you, do it." Now there were set there six waterpots of stone, according to the manner of purification of the Jews, containing twenty or thirty gallons apiece. Jesus said to

The Wedding at Cana

them, "Fill the waterpots with water." And they filled them up to the brim. And He said to them, "Draw some out now, and take it to the master of the feast." And they took it. When the master of the feast had tasted the water that was made wine, and did not know where it came from (but the servants who had drawn the water knew), the master of the feast called the bridegroom. And he said to him, "Every man at the beginning sets out the good wine, and when the guests have well drunk, then the inferior. You have kept the good wine until now!" (John 2:1-10)

The Water Molecule

He sends the springs into the valleys;
They flow among the hills.
They give drink to every beast of the field;
The wild donkeys quench their thirst.
(Psalm 104:10-11)

Molecules are super tiny. If you take a drop of water, and break it apart a billion times you would finally come down to a single water molecule. If you broke this one molecule apart you wouldn't have water anymore. You would have hydrogen and oxygen—two different substances. Water is a **molecule** made up of two hydrogen atoms and one oxygen atom, usually referred to as H_2O. When these atoms come together, they make something very different from the hydrogen and oxygen gases that make up water. Water is a **compound** (and a molecule) because it is made out of different elements. Similarly, the water molecule pictured shows the covalent bond of the hydrogen electron with the oxygen electrons.

How Would You Make a Batch of Water?

Have you ever thought about making water? If water is made out of hydrogen and oxygen, is it possible to mix up a batch of water? We don't make water like we make smoothies or trail mix. But, you can make the water compound using a **chemical process**.

Making water is a dangerous process. With professional supervision, it can be done. What makes it so dangerous is that you have to add a spark to the chemical process, which can produce fire or explosions. In other words, you must apply heat energy to disrupt the hydrogen and oxygen molecules, when you mix them together.

GOD MADE THE WORLD

Water Molecule

Oxygen Atom

Electrons

Covalent Bond

Hydrogen Atom

Hydrogen Atom

Covalent Bond

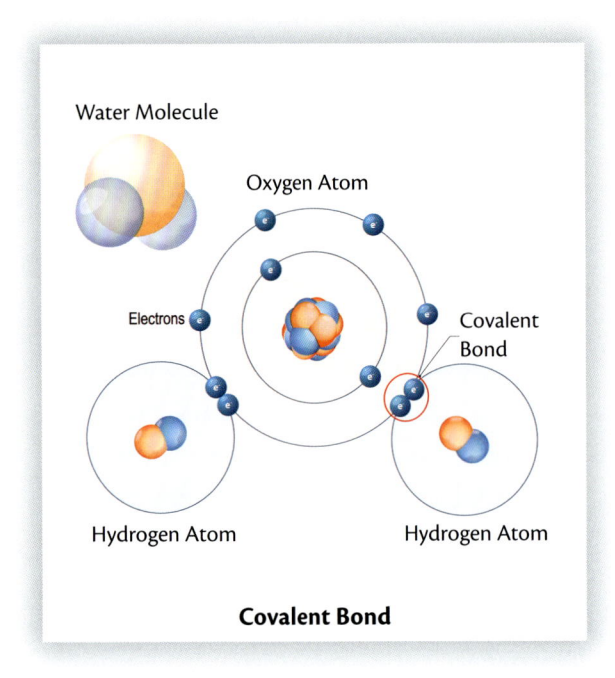

The hydrogen molecule (H₂) is nicely bonded together. So is the oxygen molecule. They are held together by what is known as a **covalent bond.** They don't want to interrupt their happy, steady state as they float around as hydrogen molecules and oxygen molecules. Similarly, the water molecule pictured shows the covalent bond of the hydrogen electrons with the oxygen electrons.

So, scientists might put some oxygen gas and some hydrogen gas into a balloon. They tie off the balloon. While shielding themselves, they touch a match or a flame to the balloon. Ka-boom! There is a violent reaction, but what drops out of the burned up balloon is...water! The hydrogen and the oxygen molecules combine to produce water.

Scientists like to describe this reaction with a **chemical equation**. Here is what it looks like:

$$2\,H_2 + O_2 \rightarrow 2\,H_2O$$

To get a whole gallon of water, you would have to blow up a balloon filled with oxygen and hydrogen about eight feet (2.4 meters) in diameter. That would be a really big balloon, and a huge explosion! Thankfully, we don't have to make our own water out of these gases. That would involve too many explosions and people would get hurt. Thanks be to God that He provided us all the water we need in the lakes, the rivers, the clouds in the skies, and the rains that water the earth!

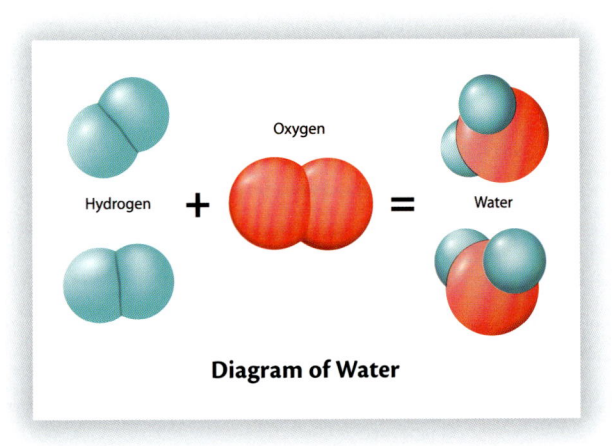

Hydrogen + Oxygen = Water

Diagram of Water

CHAPTER 4: GOD MADE MATTER

> Are there any among the idols of the nations that can cause rain?
> Or can the heavens give showers?
> Are You not He, O LORD our God?
> Therefore we will wait for You,
> Since You have made all these.
> (Jeremiah 14:22)

How to Measure Matter

> And they shall make an ark of acacia wood; two and a half cubits shall be its length, a cubit and a half its width, and a cubit and a half its height. And you shall overlay it with pure gold, inside and out you shall overlay it, and shall make on it a molding of gold all around. (Exodus 25:10-11)

Measurements are very important to God. When the Lord appointed the building of the tabernacle and the ark of the covenant, He was very specific about the widths and lengths of it all. When you buy apples or tomatoes at the grocery store, God's law specifies that the scales should be accurate. If the customer thinks he is buying five pounds of apples, the storekeeper must make sure the scale is weighing out exactly five pounds.

> Diverse weights, and diverse measures, both of them are alike abomination to the LORD. (Proverbs 20:10)

There are different ways to measure matter. We can measure all sorts of stuff in all sorts of ways.

If you are going to make cookies and cakes, you have to measure out the ingredients. If you put too much salt into the cookies, they will taste awful! How would you measure all the things we work with in the world?

1. Volume. Usually we measure volume. When you make cookies and cakes, you might add milk or water. We usually don't weigh out liquids. We usually measure it by cups, liters, milliliters, ounces, or gallons. Volume measures how much space the liquids will fill.

We don't measure the volume of solids as much, because it's a little harder to do.

You can measure a square cube or rectangular solid that has a uniform length, width, and height, this way:

Volume = Length x Width x Height

Another way to measure the volume of solids is to take a measuring cup and fill it half-way with water. Drop the solids (maybe a few stones) into the water, and see

125

how much volume was added. That's the amount of volume taken up by the solids.

2. Mass and Weight. Compare a pillow and a brick. Which has more volume? Which is heavier? The brick is heavier because it is more dense. It has more stuff in it.

Mass is not the same thing as weight. You would know the amount of mass by the number of neutrons and protons in every atom in the object. The mass is dependent on the density of the material. Feathery pillows are not dense. They are pretty light for their size. Bricks, stones, steel, and iron are quite a bit more dense than pillows.

The density of a material is related to the following two things:
- how many neutrons and protons are in the atom of the element, and,
- how crowded these neutrons and protons are spaced in the nucleus of the atom.

If there are a lot of neutrons and protons in the atom, and they are stuffed in there pretty tight, the materials will be dense. This means a little bit of the material will weigh a lot. For example, a block of balsa wood weighs a lot less than a block of iron the same size.

Actually, the most dense material on earth is osmium. It is twice as dense as lead. The chart on page 124 lists the density of different material we find on the earth.

Weight is different from mass in this way. Weight depends on how much force gravity is applying on the object. Of course, the higher the mass, the higher the force gravity plays on the object. The more mass is there, the higher the weight will be on the object.

As we already pointed out in chapter two, a 100-pound boy steps on a scale on the earth and he weighs 100 pounds. If he weighs

Pillow and Brick— Different Densities

CHAPTER 4: GOD MADE MATTER

himself on the same scale on the moon, he will only weigh 22 pounds. That's because gravity is not pulling so hard on him when he weighs himself on the moon. However, he has the same mass. His body is still made up of the same number of protons and neutrons. His body maintains the same density on the moon as it does on the earth.

3. **Hardness.** Some solid materials are very soft, and others are very hard. God made a variety of materials for different uses. If everything was made out of feathers, we could never build houses and cars. If everything was made out of rock and iron, where would we find a soft pillow for our heads?

Wood for carpentry comes in softwood (e.g. pine) or hardwood (e.g. oak). Hardwood can withstand a lot of rough handling. If

Hardwood Floor

you have a little rock in the tread of your tennis shoe and you step on a floor made out of pinewood, it will leave a deep mark in the floor. If the floor was made of oak, the little rock would not leave much of a scratch or dent.

Hardness is measured on different scales—sometimes called **Rockwell Hardness** or **Moh's Hardness**. The

GOD MADE THE WORLD

Density of Various Materials – From Lightest to Heaviest

Material	Density	Uses
Cork	0.2 gr/cm^3	sealing bottles
Pine Wood	0.7 gr/cm^3	houses, furniture
Ice	0.92 gr/cm^3	refrigeration, cold drinks
Water	1.0 gr/cm^3	drinking
Silver	10.49 gr/cm^3	money, jewelry, dentistry, photography
Lead	11.34 gr/cm^3	electrodes in batteries
Palladium	12.16 gr/cm^3	catalytic converters; watch-making, production of surgical instruments
Mercury	13.59 gr/cm^3	thermometers, electronic applications
Uranium	18.90 gr/cm^3	fuel for nuclear power plants
Gold	19.32 gr/cm^3	money, investment, jewelry
Tungsten / Wolfram	19.60 gr/cm^3	heavy metal alloys and steels
Plutonium	19.80 gr/cm^3	nuclear weapons; electrical power generation
Platinum	21.40 gr/cm^3	catalyst in chemical reactions
Osmium	22.61 gr/cm^3	alloyed with other metals, fountain pens

Note: A cubic centimeter is the same as a milliliter.

following table compares the hardness of different kinds of material.

Material	Moh's Hardness
Chalk	0.8
Fingernail	2.2
Gold	2.5
Your Tooth	5.0
Knife Steel	5.5
Glass	6.2
Tungsten Carbide	9.0
Diamond	10.0

Hardness makes a difference if you want to cut things up. You can't cut a knife blade with your fingernail, but you can cut your fingernail with a knife. You can bite into a gold coin and make a dent in it, because your teeth are twice as hard as gold. Circular saw blades are coated with diamond tips so as to cut through really hard stuff like bricks. If you want to cut

Cutting Bricks

through sheet metal, you can't use a really soft blade. You want to use a Tungsten Carbide-tipped blade or a diamond-tipped blade to do the job. These are much harder than regular steel blades. They will chop right through the sheet metal.

Softwoods are best used for lumber to make houses, as well as for paper and cardboard boxes.

Hardwoods are best used for making

Paper

Drilling Through Sheet Metal

wood floors, baseball bats, and handles for axes and shovels.

Wood has a lot of benefits. It's easy to find. God grows it everywhere, all over the world. You don't have to dig wood out of the ground. Wood is easy to cut with saws. You can pound nails through it and fasten it easily. It looks beautiful when it is finished with oils and sealant. Without wood, we would have a hard time constructing buildings, homes, furniture, cabinets, and tables. What a blessing that God would grow all that wood all over the world, so convenient for our use! Can you believe there are about 400 trees for every human in the world? That's enough wood to build about 20 houses for each person on earth.

Nothing is wasted from this lumber. Sawdust is used to make glue. And bark can be used to tan leather, as well as make

Rope—Lower Tensile Strength Than Steel

cork board, and mulch for gardens. Paper is made from pulpwood, which is finely chopped wood particles.

4. Tensile Strength. Some solid materials are stronger than others. The tensile strength is the amount of strength the material withstands if you were to pull it apart. The strength of a material is measured in mega-pascals (Mpa) or pounds per square inch (psi). The strongest material in the world is graphene, made out of carbon. It is used to make cellphone screens, tennis rackets, automobiles, and high-end bicycle frames. Suppose that you had a thin strand of each of the following materials, sized as tiny as a human hair. How much weight could the material hold before breaking? This is called the **tensile strength** of the material.

God made some material very strong so it could be used to hold up under lots of stress. For example, saw blades are sometimes coated with diamond so as to cut through really hard things like bricks.

High Tensile Strength Steel Wire

GOD MADE THE WORLD

Metal

Surely there is a mine for silver,
And a place where gold is refined.
Iron is taken from the earth,
And copper is smelted from ore...
But where can wisdom be found?
And where is the place of understanding?
Man does not know its value,
Nor is it found in the land of the living.""
(Job 28:1-2,12-13)

Besides wood, the other solid material which God gave us to use is metal. Most metal is made out of a combination of materials we get out of the earth. Metal is used for building cars, ships, bridges, steel buildings, electronics, stoves, and refrigerators.

The most commonly used material in the world is crude steel—a total of 1.8 billion tons of the metal is made every year. China, India, Japan, and the United States are the largest producers in the world. Steel is made from iron ore that God put in the ground for our use when He made the world. This iron ore has a lot of impurities in it—things like silica, phosphorus, and sulfur. So steelmakers will heat up the ore to very hot temperatures, until it reaches a liquid state. Then, oxygen is blown into the hot ore. The oxygen gas combines with the impure elements to make a slag, which bubbles to the surface of the mix. The slag is skimmed off. Then, the steelmakers will mix carbon into the hot iron ore to give strength to the steel. Chromium is added to make **stainless steel**. This keeps the steel from getting rusty.

Copper is used for

Boeing 747 Jet, Built with 100 Tons of Aluminum

CHAPTER 4: GOD MADE MATTER

Most Common Raw Building Material	Annual World Consumption
Iron Ore	2 billion tons
Metallurgical Coal	1 billion tons
Wood for Building	600 million tons
Wood for Paper	400 million tons
Aluminum	60 million tons
Glass	6.2 million tons
Tungsten Carbide	9.0 million tons
Copper	23.8 million tons

fairly easily recycled without spending too much money doing it. The processing of aluminum begins with bauxite, a sedimentary rock mined mainly in South America or Australia. This rock is ground up very fine and mixed with a little bit of water which makes a paste. This paste is heated with steam, which removes the silicon. The problem is that this is an aluminum oxide compound. So the mixture is mixed in with molten cryolite and heated to 950 C. An electric current of 400,000 Amps is introduced to the mix, and that breaks the oxygen bond and the pure aluminum settles at the bottom of the vat.

The inventor of aluminum processing was Charles Martin Hall, a electric wire and electric motors.

Aluminum is excellent for many uses. It is lightweight and it doesn't rust. So it is terrific for roofs and pans, wherever water will be present. Aircraft manufacturing uses a lot of aluminum. A Boeing 747 jet uses 100 tons of the metal. It can be

Young Charles Martin Hall

God-fearing man, and the son of a Presbyterian missionary in Jamaica. Homeschooled by his mother, he invented aluminum using his family's furnace at their home in 1886 with a little help from his sister.

*For the LORD gives wisdom;
From His mouth come knowledge and understanding;
He stores up sound wisdom for the upright;
He is a shield to those who walk uprightly.
(Proverbs 2:6-7)*

Cement Mixture

Concrete

Don't forget the blessing of concrete when you think of important mixtures used to help mankind. Look around you, and you will find concrete in building foundations, roads, driveways, dams, skyscrapers, sidewalks, and bridges.

Concrete is a mixture of cement, sand, small rocks, pebbles, and water. Cement is a binder that can harden around the rocks and pebbles, creating a super strong road base or building foundation. The cement mixture includes crushed up limestone, silica, alumina, and other chemicals. When you add water to the cement, it hardens through a chemical process called **hydration**.

For a really strong concrete, builders will mix up about 10-15% cement, 60-75% small rocks and pebbles, and 15-20% water. Interstate highways require about 11-12 inches of concrete. This becomes a strong enough base for huge 18-wheel-

CHAPTER 4: GOD MADE MATTER

Material	Tensile Strength	Weight Held by a Hair-Size Strand
Human Bones	130 MPa	2 ounces
Silicone Carbide	138 Mpa	2 ounces
Human Hair	196 MPa	3 ounces
Titanium Alloy	1000 Mpa	15 ounces
Spider's Silk	1000 Mpa	15 ounces
Tungsten Carbide	1510 Mpa	23 ounces
Strongest Steel	2693 Mpa	40 ounces
Diamond	2820 Mpa	43 ounces
Glass Fiber	3450 Mpa	52 ounces
Carbon Nanotubes	33,000 Mpa	440 ounces
Graphene	130,000 Mpa	1,950 ounces (a 125-pound boy)

er trucks to roll over day after day. The toughest roads in the world are made out of 24 inches of gravel, 6 inches of asphalt, and 14 inches of concrete on top of it all. That's almost 4 feet of material to allow for 300,000 cars and trucks to roll over them every single day!

GOD MADE THE WORLD

Cement Mixture	
Lime	60-65%
Silica	17-25%
Alumina	3-8%
Magnesia	1-3%
Iron Oxide	0.5-6%
Calcium Sulfate	0.1-0.5%
Sulfur Trioxide	1-3%
Alkaline	0-1%

God Wants Us to Make Use of His Creation

"When [the LORD] prepared the heavens, I [Wisdom] was there,
When He drew a circle on the face of the deep,
When He established the clouds above,
When He strengthened the fountains of the deep,
When He assigned to the sea its limit,
So that the waters would not transgress His command, when He marked out the foundations of the earth,
Then I was beside Him as a master craftsman;
And I was daily His delight,
Rejoicing always before Him,
Rejoicing in His inhabited world,
And my delight was with the sons of men."
(Proverbs 8:27-31)

What an amazing world God has created for us! Surely, the whole creation is a testimony to His matchless wisdom. Let us consider also the goodness of God. There are millions of useful materials all around us. There is so much of God's wise provision yet for us to discover. In His mercy and wisdom, the Lord has provided a world of useful chemicals and materials to assist life here on this earth.

God used Robert Boyle, Roy Plunkett, and Charles Martin Hall to bring about extremely useful discoveries like Teflon, aluminum, and the field of chemistry.

We need more godly scientists to explore God's good earth with this purpose in mind. Could you find some wonderful discovery that will help your fellow men? There is much left to discover.

CHAPTER 4: GOD MADE MATTER

Robert Boyle challenged his fellow Christians to be heroes in this important scientific work. Science is honorable work.

There are no men that seem to me to have nobler and sublimer aims, than those to which a true Christian is encouraged; since he aspires to no less things than to please and glorify God; to promote the good of mankind and to improve, as far as is possible, his personal excellencies in this life; and to secure to himself for ever a glorious and happy condition in the next.

Excavator at a Mine

GOD MADE THE WORLD

Pray

- Praise the Lord for His unsearchable wisdom in creating the atom, the electrons, the protons, the neutrons, and the mysterious forces that keep the atom together.
- Thank God for all of the elements He has made, and the usefulness of His creation. Thank Him for water, wood, air, aluminum, Teflon, chemical reactions, mixtures, soft drinks, cement, and steel.
- Pray for wisdom to make new discoveries about God's world.

Sing

For the Beauty of the Earth

For the beauty of the earth
For the glory of the skies,
For the love which from our birth
Over and around us lies.

Chorus: LORD of all, to Thee we raise,
This our hymn of grateful praise.

For the beauty of each hour,
Of the day and of the night,
Hill and vale, and tree and flower,
Sun and moon, and stars of light.

For the joy of ear and eye,
For the heart and mind's delight,
For the mystic harmony
Linking sense to sound and sight.

For the joy of human love,
Brother, sister, parent, child,
Friends on earth and friends above,
For all gentle thoughts and mild.

For Thy Church, that evermore
Lifte'th holy hands above,
Offering up on every shore
Her pure sacrifice of love.

For each perfect gift of Thine,
To our race so freely given,
Graces human and divine,
Flowers of earth and buds of Heaven.

If you do not know the hymn, you may listen to a version of the hymn on the Internet, with supervision, and sing along with it.

Do

Choose one of the following activities and apply the lessons you have learned in this chapter.

1. **Invent.** Try to create a new smoothie mixture that has never been tried before. The healthiest vegetables are usually spinach, kale, carrots, broccoli, garlic, green peas, Swiss chard, ginger, asparagus, and red cabbage. The healthiest fruits are usually grapefruit, pineapple, avocado, blueberries, apples, pomegranate, mango, strawberries, cranberries, lemons, and dorian. Select the vegetables and fruits that would provide the best balance of nutrients, vitamins, minerals, and antioxidants. Experiment with various amounts of each of the substances you choose. Try adding a few spices. Be sure your new mixture tastes good. Consider marketing your new mix. (For this exercise, we recommend parental or adult supervision. You will need to use a blender for this exercise.)
2. **Solve a problem.** Identify a problem in the home, and come up with a solution to the problem.

Watch

To watch the recommended videos for this chapter, go to **generations.org/GodMadeTheWorld** and scroll down until you find the video links for Chapter 4. Our editors have been careful to avoid films with references to evolution. However, we would still encourage parents or teachers to provide oversight for all internet usage. These videos may not give God the glory for His amazing creative work, so the student and parent/teacher should respond to these insights with prayer and praise.

Wind Turbines

Chapter 5

GOD GIVES US ENERGY

> By awesome deeds in righteousness You will answer us, O God of our salvation, You who are the confidence of all the ends of the earth, and of the far-off seas; Who established the mountains by His strength, being clothed with power; You who still the noise of the seas, the noise of their waves, and the tumult of the peoples. (Psalm 65:5-7)

There are two big things that God has given to us in this universe He has made. In the last chapter, we talked about the first thing God gave us—matter or materials, the stuff you can see all around you. Now, the other important thing that God gave to us when He created the world is **energy**. **Physics** is the study of matter and energy.

God Gives You Energy

This morning you were lying in bed hardly moving at all. Then, you climbed out of bed and made your way to the bathroom. How did you get the energy to do this? Moving a body from here to there takes energy. Where did that energy come from?

The energy propelling your body comes from food that is processed into energy. The food is carried to the cells by the bloodstream, and then the cells in your body process the food into energy.

GOD MADE THE WORLD

Have you not known?
Have you not heard?
The everlasting God, the LORD,
The Creator of the ends of the earth,
Neither faints nor is weary.
His understanding is unsearchable.
He gives power to the weak,
And to those who have no might He increases strength.
Even the youths shall faint and be weary,
And the young men shall utterly fall,
But those who wait on the LORD
Shall renew their strength;
They shall mount up with wings like eagles,
They shall run and not be weary,
They shall walk and not faint. (Isaiah 40:28-31)

God is Limitless in Energy

Sometimes we can get very tired. After climbing a mountain or working hard all day, we run out of energy. Eating food helps to renew that energy. But, you need to know that God never runs out of energy. He doesn't even need to sleep. Psalm 121:4 tells us that "He who keeps Israel shall neither slumber nor sleep." Indeed, the Lord our God is the very source of all of the energy and power in the universe!

Ah, Lord GOD! Behold, You have made the heavens and the earth by Your great power and outstretched arm. There is nothing too hard for You. (Jeremiah 32:17)

Moving Heavy Objects is Work

Energy is the ability to do work or to create heat. Work is done when forces are pushing on things to move them around. When you push a piano across the living room floor you have done work. When you give a shove

Slingshot Pulled Back—Potential Energy

to a little toy car which is sitting still, and it starts to move, you have done work. A child lying in bed isn't doing any work. Work happens when things begin to move.

What if God did not give us energy? We would never be able to do anything. We would not be able to work. The whole world would be static. There would be no change in anything. There would be no movement. It would be a very boring world. God wanted a world where there would be change going on, so He gave us energy.

There are two kinds of energy—kinetic and potential energy. **Potential energy** is the potential to do work. Think of a wound-up spring all ready to spring out. The spring has great potential to do work, expend energy, or engage an action. **Kinetic energy** is the energy being used when something is in motion. Once you let the wound-up spring go, the potential energy turns into kinetic energy.

Kinetic and potential energies come in different forms including mechanical, sound, thermal, electric, magnetic, gravitational, chemical, ionization, nuclear, and elastic. You don't need to know about all of these forms, except that God made them all. We will explore some of these energy forms in this chapter. The following table gives some examples of how these energies are used in our world.

GOD MADE THE WORLD

Energy Type	Kinetic or Potential	Examples
Mechanical	Kinetic	Moving wind energy, muscles, machines
Thermal	Kinetic	Geothermal
Electrical	Kinetic	Lightning, household current
Electro-magnetic	Kinetic	Microwave ovens, radio, light, solar
Sound	Kinetic	Voices, thunder, automobile engine noise
Gravitational	Potential	Non-moving waterwheel, roller coaster
Nuclear	Potential/Kinetic	Atom/nuclear Power
Magnetic	Potential	Magnets
Elastic	Potential	Stretched rubber bands and springs

Energy Forms Can Change into Other Energy Forms

In order to make energy more useful to us, God made sure that one energy form can change into another. Batteries produce energy by a chemical reaction. If you put batteries into a flashlight, the chemical energy turns into electrical energy. The electrical energy lights up a filament turning into heat, which gives off a helpful light to the camper who is trying to pitch his tent in the dark.

God's System of Laws

God set up the world according to a system of natural laws. He wanted the world to run in the same way all the time. Scientists have discovered many laws by which God runs the world.

When it comes to observing God's

CHAPTER 5: GOD GIVES US ENERGY

world and how it works, scientists try to explain it in different ways. Good scientists want to be humble and truthful. They present their discoveries using hypothesis, theory, and law.

Forces and Kinetic Energy

You may remember that the devout Christian Robert Boyle was the most important scientist in the field of chemistry. God used another thoughtful man to introduce modern physics to the world—Isaac Newton. As this brilliant

Sir Isaac Newton (1643-1727)

> **Hypothesis**—An educated guess for which scientists don't have much evidence. Good scientists will say they are **not very sure** about a hypothesis.
>
> **Theory**—A conclusion made after testing a hypothesis. Good scientists will say that they are **kind of sure** about their theories.
>
> **Law**—A firm conclusion about the world. Scientists will tell us that they are **very sure** about a law of nature. This comes by a great deal of observation and testing.

scientist studied this amazing creation, he finally concluded:

> This most beautiful system of the sun, planets and comets, could only proceed from the counsel and dominion of an intelligent and powerful Being. . . this Being governs all things, not as the soul of the world, but as LORD over all; and on account of his dominion he is wont, to be called Lord God. . . or Universal Ruler.[1]

Isaac Newton's First Law of Motion is this: A body will remain at rest or in

Coasting on Bikes

uniform motion on a straight line until an external **force** acts on it.

A force is not the same thing as kinetic energy. You need a force to get something moving, or to get it to speed up. Kinetic energy is the energy of a mass that is in motion. It is a force that will give some thing kinetic energy. For a car to double its speed, it needs a little shove. The car needs a force applied to it.

Have you ever been riding your bike along on flat ground, and you stopped peddling for a while? Your body has stopped applying a force to the bike. No extra energy has been added while you are coasting on flat ground.

Newton said that a body in motion will keep moving, unless some force slows it down. But what about your bike? If you were coasting along on a flat surface, would your bike slow down and stop after awhile? What is it that makes the bike slow down if you are not applying the brakes? It turns out that the dirt and the rocks rubbing against your bike tires are slowing you down. This **friction** is applying a force against you to slow you down. Also, you can feel the air or the wind blowing against you as you coast along. This is called **air resistance**. When bikes and cars are moving along on the earth, air resistance and friction from the dirt or pavement will slow them down.

Now, suppose you were riding in a rocket ship through space. There would be no dirt, rocks, or air resistance to slow you down. Suppose your rocket would speed your space ship up to 400,000 miles per hour, and then the space capsule starts to coast. Would you ever slow down? You would not. There would be no wind resistance pushing against you. Space astronauts have tested Newton's theory. Sure enough! A space ship remains in uniform motion on a straight line until an external force acts on it.

Matter Can Be Changed Into Energy

God created both matter and energy. Here's something neat about God's created world: you can exchange matter for energy and energy for matter!

Where does a car's energy come from? How do you get a car going from 0 mph to 75 mph? The car's engine burns gasoline which turns heat energy into mechanical energy. We will explain more of how an engine works in the next chapter.

What you need to know is that gasoline is matter—a liquid that comes from oil. When the gasoline is burned in the engine, a tiny bit of matter actually turns into energy. A scientist by the name of Albert Einstein related matter to energy by this formula:

$$E = mc^2$$

or

Energy = mass x (the speed of light) x (the speed of light)

Basically, what this formula tells us is that a tiny bit of mass can create a huge amount of energy. If a mass the size of a 100-pound rock were turned into energy, that mass

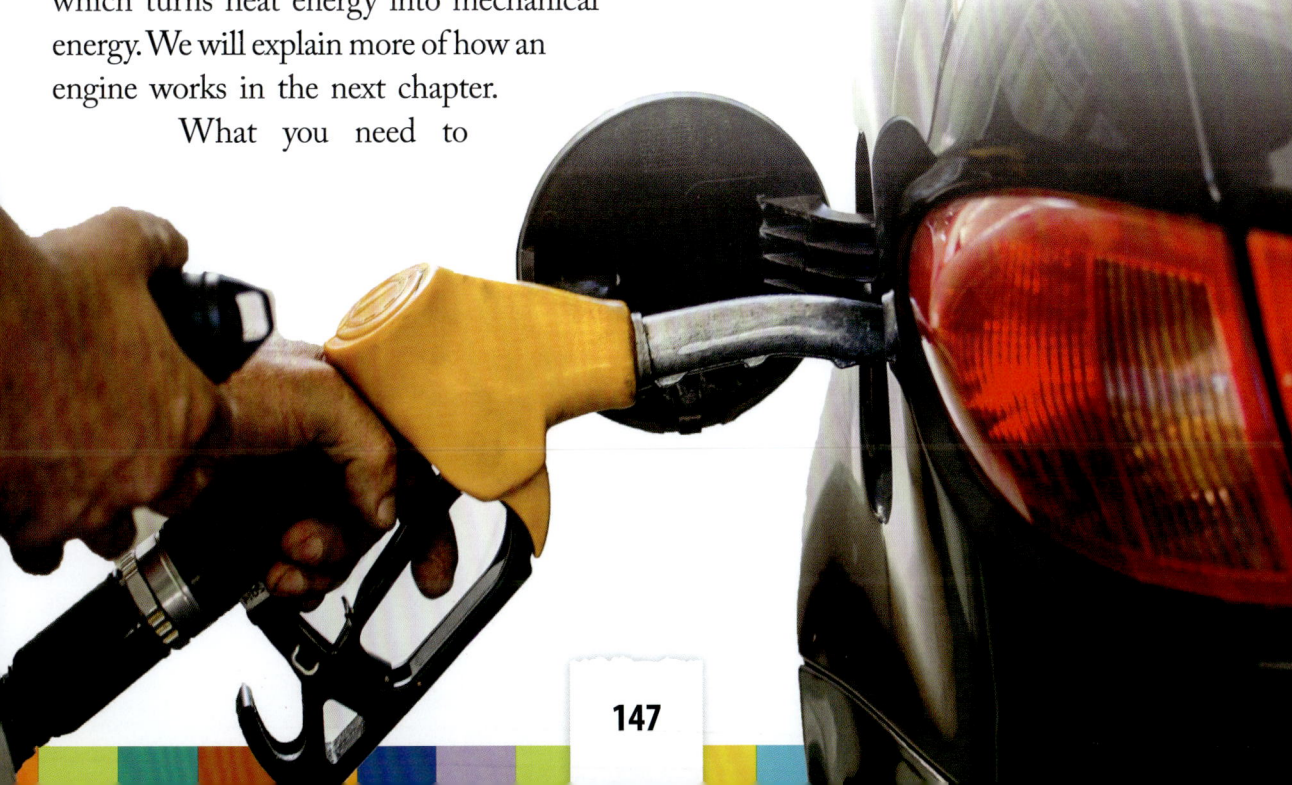

GOD MADE THE WORLD

would produce 4.5 x10^{18} joules. That would be equal to the energy of 80,000,000 nuclear bombs! This is also one reason why a sun could burn for billions of years without running out of mass. The sun is a nuclear fusion reactor, where atoms are constantly turning into energy—heat and light. Scientists believe they have found ways to turn energy back into matter.

Heat Energy

Heat is an important kind of energy used to warm up homes, power car engines, and cook food.

To get a car going, heat energy is transferred into mechanical energy or mechanical movement. When a vehicle begins to move, we say that it has added mechanical energy. First, the fuel injectors squirt a little gasoline into one of the cylinders. A spark lights off the gasoline which creates more heat and pressure in the cylinder. This causes the piston to move up and down in the cylinder. This movement turns the camshaft which rotates the drive shaft and turns the wheels.

What produces heat? What makes some things hotter than others? The heat in a substance is really the total activity of

Heat Energy Transfers to Motion in Car Engine

CHAPTER 5: GOD GIVES US ENERGY

Temperature	Kelvin	Celsius	Fahrenheit
Absolute Zero	0° K	-273° C	-460° F
Water Freezes	273° K	0° C	32° F
Water Boils	373° K	100° C	212° F

all the atoms and molecules. The faster all the particles move, the hotter the stuff will be. **Thermal energy** is the total kinetic energy of the atoms and molecules in the substance. A cup of hot water has more thermal energy than a cup of cold water. But a huge barrel of cold water could have more thermal energy than a little cup of hot water. That's because there is more mass in the barrel of cold water. And, even though the water is colder there is still some movement of the molecules in the barrel.

Temperature is the measurement we use to determine how much thermal energy is in a substance. Some of the little particles may be moving faster than others. A thermometer measures the average heat energy in the surrounding substance.

There are three scales used to measure temperature: Fahrenheit, Celsius (or Centigrade), and Kelvin. Temperatures can get very high as materials get hotter. Scientists have been able to get a substance up to 5.5 trillion degrees Kelvin. However, materials cannot get any colder than -273°C, or absolute zero. At this temperature, there is a total absence of heat energy.

A Christian man named William Thomson (also known as Lord Kelvin) developed the Kelvin scale. He was the primary scientist who developed the study of heat—called **thermodynamics**. The three scales are compared above.

One hundred years ago, people would use bricks heated up in the fireplace to keep themselves warm. They would bring the bricks to church to keep the family

GOD MADE THE WORLD

warm as they sat in the pews. They would take a brick to bed with them as well. How long do you think the heat from the brick would keep a child warm in bed? Well, that depends on two things—the size or weight of the brick, and the temperature of the brick. Suppose you were taking a brick to bed with you to keep you warm on a cold night. You had a choice of two bricks—a two-pound brick at 150°F or a one-pound brick at 180°F. Do you see how the bigger brick at the lower temperature would give off more heat throughout the night? You could test this by heating up

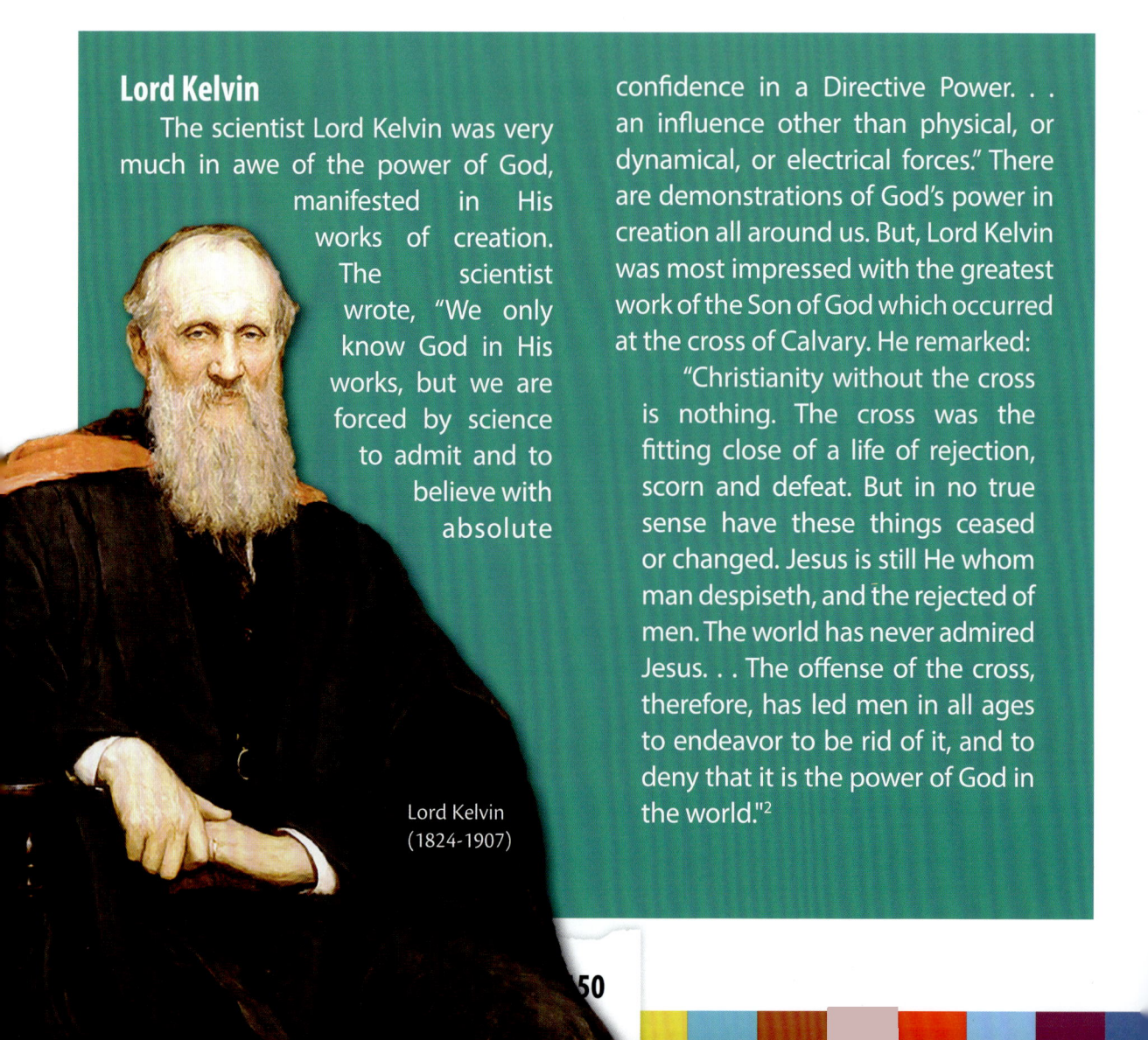

Lord Kelvin

The scientist Lord Kelvin was very much in awe of the power of God, manifested in His works of creation. The scientist wrote, "We only know God in His works, but we are forced by science to admit and to believe with absolute confidence in a Directive Power. . . an influence other than physical, or dynamical, or electrical forces." There are demonstrations of God's power in creation all around us. But, Lord Kelvin was most impressed with the greatest work of the Son of God which occurred at the cross of Calvary. He remarked:

"Christianity without the cross is nothing. The cross was the fitting close of a life of rejection, scorn and defeat. But in no true sense have these things ceased or changed. Jesus is still He whom man despiseth, and the rejected of men. The world has never admired Jesus. . . The offense of the cross, therefore, has led men in all ages to endeavor to be rid of it, and to deny that it is the power of God in the world."[2]

Lord Kelvin
(1824-1907)

CHAPTER 5: GOD GIVES US ENERGY

two rocks in boiling water—one weighing one pound and the other weighing two pounds. Put both rocks in a cup of cold water and measure the temperature of the water after a few minutes.

Expansion and Contraction

A thermometer works by a principle of **expansion**. When a substance gets hotter, it usually expands. As the particles move faster and faster, the material expands. Thus, when the mercury (or other liquid material) in a thermometer gets hotter the liquid expands. If the liquid heats up only about 5 degrees, it doesn't expand very much. Thermometers are made with a bulb of liquid feeding up into a very small cylinder. It doesn't take much expansion for the liquid to rise in the cylinder. So with small temperature differences, you will see the liquid rise. Mercury works best for thermometers because it expands more than other liquids. Yet, Mercury can also be dangerous if it is taken into your body.

Water also expands when it is heated up. Suppose you heated up a five-foot-deep swimming pool from 80°F (27°C) to 210°F (99°C). The depth would increase by three inches.

Solids also expand when they are

Gaps in Sidewalk Allow for Contracting and Expanding

heated. Roads and bridges can break apart with the changing seasons. In the winter time, pavement will **contract** and in the summer time, it will **expand**. When engineers lay out the bridges, parking lots, sidewalks, and streets, they usually put regularly-spaced cracks in the pavement. This allows for expansion and contraction throughout the year. If they didn't include these spaced gaps, the concrete would contract and buckle in the winter. And, it would expand and create more cracks in the summer time.

Heat Transfer

The really interesting thing about heat is that it tends to move around. Just like water flows from here to there, heat flows too. Just as light flows into darkness,

151

and the darkness does not flow into the light, heat energy contained in the hotter substance flows into the colder substance and makes it hotter. This is called **heat transfer.** Our Creator has provided three ways for heat to move from here to there—by **conduction, convection,** and **radiation**.

Conduction and Convection

You need to know the difference between conduction and convection. Conduction happens when excited atoms get other atoms excited by colliding against each other. Convection happens when the excited atoms jump around and actually move from one place to another. As the atoms get hotter, they spread out even more. That's why convection usually happens in liquids and gases. Conduction usually happens in solids. Heat will also conduct from liquids into solids.

You can get convection going when you open the front door of your house on a cold day. The hot air from inside tries to escape to the outside. The excited atoms in the hot room inside exit through the doorway into the cold outside.

You experience conduction when you put your hand on a hot stove. The hot and excited atoms try to get the atoms in

Wooden Spoon in Pot

your hand hot and excited as well. If you were to hold on to a hot surface that was 160°F (70°C) for less than a second, your skin would sustain a serious burn. The hot atoms are exciting the atoms in your skin—and raising the temperature. This is an example of conduction between a solid and a solid.

Heat conducts from a liquid into a solid when you stir hot soup using a steel spoon. The heat in the liquid will excite the atoms on the spoon that is submerged in the liquid. But, after awhile the atoms on the lower part of the spoon will excite the atoms further up the spoon. Gradually, the heat will crawl up the spoon. The spoon will eventually become just as hot as the boiling soup—about 212°F (100 °C). If you grab the spoon, you will burn your

CHAPTER 5: GOD GIVES US ENERGY

hand, as the heat on the spoon conducts to your hand.

Some spoons conduct heat better than others. They are called **conductors**. Usually, moms will use wood spoons or plastic-handled spoons to stir hot food on the stove. This protects their hands from getting burned. Wood and plastic are called **insulators**, because they do not conduct heat. The following table shows the best insulators and the best conductors by a rating number.

Air is the best insulator. That's why birds will fluff out their feathers to keep warm on a cold day. God has given them this way to provide an extra layer of insulation when they need it.

Firewalking

Firewalking is one of those natural mysteries that is rather easily solved. For thousands of years people have performed the risky feat of walking over hot coals in their bare feet. It's a dangerous business. Some people get burned and some don't. Walking across coals without getting burned has to do with conduction. From the table on the right, you can see that wood, skin, and water are poor conductors of heat. The walkers dip their feet into water first, which is also an insulator. Then

Conductor Material	Thermal Conductivity Rating
Diamond	1000
Silver	429
Copper	386
Aluminum	204
Cast Iron	73
Stainless Steel	16

Insulators	Thermal Conductivity Rating
Glass	1.05
Concrete	1.00
Brick	0.69
Water	0.60
Plastic	0.42
Skin	0.30
Wood	0.1
Drywall	0.05
Fiberglass	0.045
Wool	0.045
Air	0.024

GOD MADE THE WORLD

they proceed to walk quickly so there isn't much time for the heat to conduct from the coals into their feet.

Radiation

Oh, give thanks to the LORD, for He is good!
For His mercy endures forever...
To Him who made great lights,
For His mercy endures forever—
The sun to rule by day,
For His mercy endures forever;
The moon and stars to rule by night,
For His mercy endures forever.
(Psalm 136:1, 7-9)

The sun is the perfect heater for this earth. It is a constant, daily reminder of the mercy of God upon us. Without the sun, we would have no warmth, no growth for plants, no rain, and no life for animal or man. Indeed, His mercies are new every morning, and there is no end to the sunshine of His love for us.

Put simply, conduction and convection happen when things come into contact with other things. A spoon has to be in contact with a liquid for heat conduction. Air in one room has to be in contact with the air in another room, so fast-moving atoms can wander into the next room by way of convection.

Now, how does the heat of the sun reach the earth if there is no gas, no air, and no fluids to bring it here? Once you get outside of the world's atmosphere, there is nothing but empty space all the way to the sun. In His great wisdom, the Lord provided radiation. At the beginning of the world, God created an energy called **electromagnetic radiation**. Technically, He created this on the first day, when He said, "Let there be light" (Genesis 1:3), and there was light.

This mysterious energy travels through space on a wave. Both light and heat come from the sun and connect to our world by **electromagnetic waves**. Once the electromagnetic energy hits the

Radiative Heater

earth—the rocks, the dirt, the water, and you—it changes into heat energy. Black surfaces especially heat up when hit by these waves, because black **absorbs** all wavelengths of light. On the other hand, white surfaces will **reflect** all the various wavelengths of light.

Radiation is all around us. The fire in your wood stove radiates heat. Toasters, x-rays, and microwave ovens use this electromagnetic radiation. Sometimes a fan will blow the heat out of a fireplace, and you can feel the result of convection. Without the fan blowing, though, you can still feel heat radiating on your body from the fire. It is a different kind of heat. It travels much faster than convection. It comes by way of an electromagnetic wave. This is radiation.

Our kind and merciful God has used Christians to bless billions of people throughout the world with key scientific breakthroughs over the last 1,000 years. The man who discovered the electromagnetic wave was another committed Christian scientist named James Clerk Maxwell. He was a man of prayer, and we read this from his notes: "Almighty God, Who hast created man in Thine own image, and made him a living soul that he might seek after Thee, and have dominion over Thy creatures, teach us to study the works of Thy hands, that we may subdue the earth to our use, and strengthen the reason for Thy service."[3] The greatest scientist of the 20th century, Albert Einstein, claimed that he stood on the shoulders of James Clerk Maxwell.

Toasters Toast Bread by Radiation

Statue of James Clerk Maxwell (1831-1879) in Edinburgh, Scotland

Getting Rid of Unnecessary Heat

Sometimes, you get heat when you don't want it. Incandescent lightbulbs produce both light and heat. Since you don't really need the heat from the lightbulb, it is a slight waste of energy. It also makes the room hotter, which you would not want on a hot summer night. This extra heat is called a **byproduct**.

The automobile burns gas which turns into mechanical energy. But, that burning also creates heat. Most vehicles have radiators and fans to cool off the engine. The radiator moves a fluid through the engine block to cool it off. If an engine overheats, the damage to the vehicle can be very expensive. You want to be sure that the fluids in your car do not run low.

If you listen carefully to your home computers, you will hear fans cooling off the electronics. The fan is blowing heat away from the computer parts by the process of convection.

Our Creator God provided some really creative means for cooling down animals and humans on hot days. Dogs pant and humans sweat. Circulating blood in the body cools off your skin. The blood carries your body heat to the surface, where the heat is convected into the surrounding air. Also, sweat glands create moisture on your skin. As air moves across your body, the moisture evaporates—creating a cooling effect on your body. This is another example of convection.

The Electromagnetic Force

"Can you search out the deep things of God?
Can you find out the limits of the Almighty?
They are higher than heaven—what can you do?
Deeper than Sheol—what can you know?
Their measure is longer than the earth
And broader than the sea." (Job 11:7-9)

In this lesson, we embark on exploring more mysteries. If you cannot understand all that God has done, don't worry about it. Most scientists cannot fully explain things like electromagnetism, the earth's magnetism, the north and south poles of magnets, and many other mysteries. God's mind is broader than the sea and larger than the universe. The main goal of all of this study is to worship God, and to stand in wonder at His amazing work.

Before discussing electrical energy, let's review the four mysterious forces that God created in the universe. They are:

CHAPTER 5: GOD GIVES US ENERGY

1. The Gravitational Force
2. The Weak Force in the Atom's Nucleus
3. The Strong Force in the Atom's Nucleus
4. The Electromagnetic Force

The electromagnetic force is the electrical charge the Lord set up between the protons and electrons in the atom. Like charges repel each other, and opposite charges attract each other. The proton was given a positive charge and the electron a negative charge. This is where electrical energy comes from.

Normally, the electrons are happy to be close to the protons in the atom. But, there are things you can do to separate the electrons from the protons. Separating a lot of electrons from the protons is what creates an electric charge.

Magnetism

Magnetism is one of the effects of the electromagnetic force. When something has become magnetized, it is called a **magnet**. Some materials experience the effect more than others.

Magnet and Steel Balls

Typically electrons will spin around the nucleus in random directions. A magnetic field is created by a bunch of rotating electric charges. This field causes a magnetization of materials. Mysteriously, in some materials, the electrons begin to spin around the nucleus in the same direction.

Once a material is magnetized, the magnet will have a north and south end to it. If the electrons are spinning counterclockwise, the south end of the magnet will position in the upwards direction. If the electrons are spinning clockwise, the north end of the magnet will position in the upward direction.

Iron, cobalt, neodymium, and nickel are the most magnetic of God's elements. Most materials will magnetize for just a short time, but some magnets are permanent. They never lose their magnetism.

Scientists found a way to make permanent magnets in the 1930s. Earlier in the history of the world, people would find "lodestones." These were rocks which were probably magnetized by a lightning strike, which caused the electrons in the material to spin the same direction.

Fisherman occasionally use very strong magnets made of neodymium to retrieve lost objects out of rivers and lakes. These magnets weighing just four pounds can pull up to 1,300 pounds of steel out of rivers and lakes.

The Earth's Magnetic Core

Every magnet has a magnetic north pole and a magnetic south pole. You can identify the north pole and the south pole by the direction the electrons are spinning.

The earth is a gigantic magnet with the magnetic south located up around the North Pole. The magnetic north is

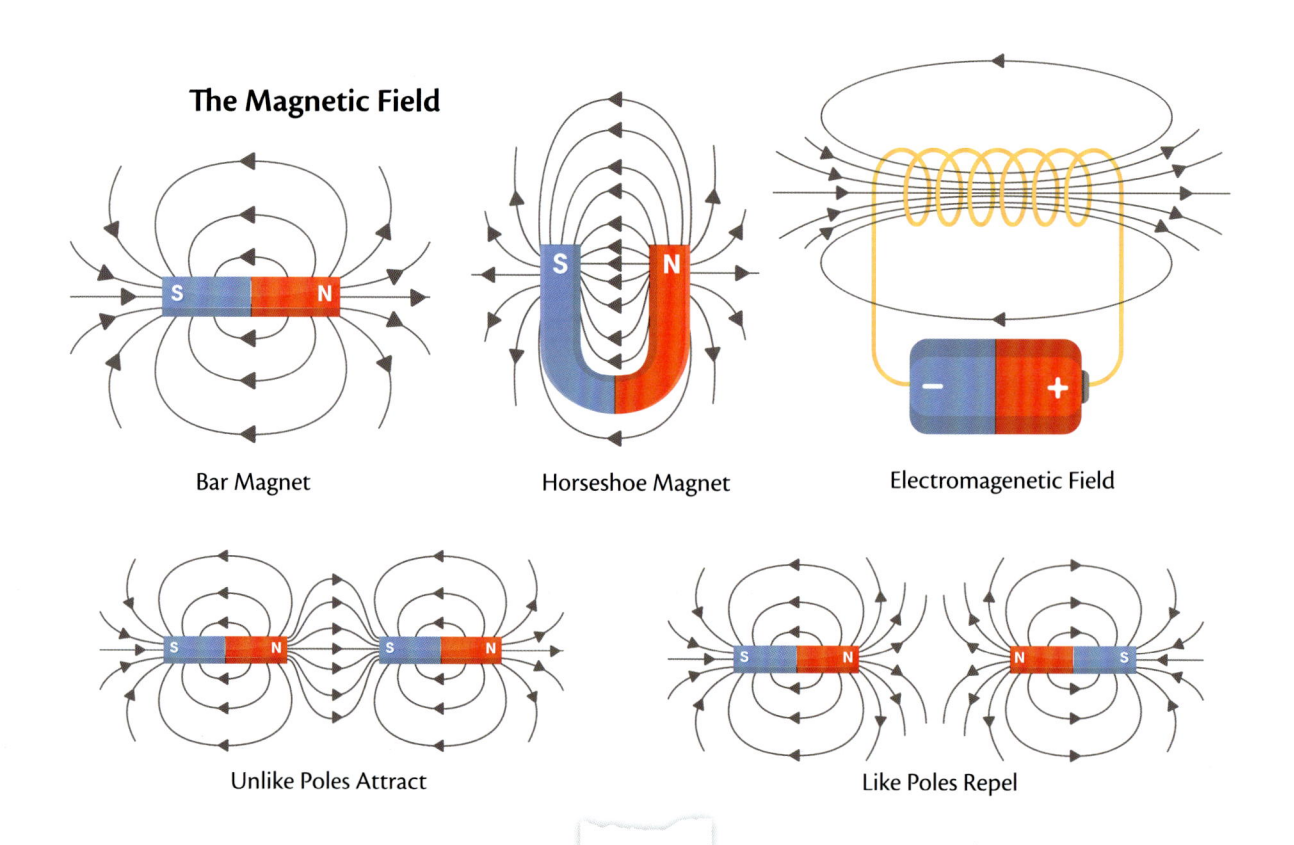

The Magnetic Field

Bar Magnet

Horseshoe Magnet

Electromagenetic Field

Unlike Poles Attract

Like Poles Repel

Earth's Magnetic Field

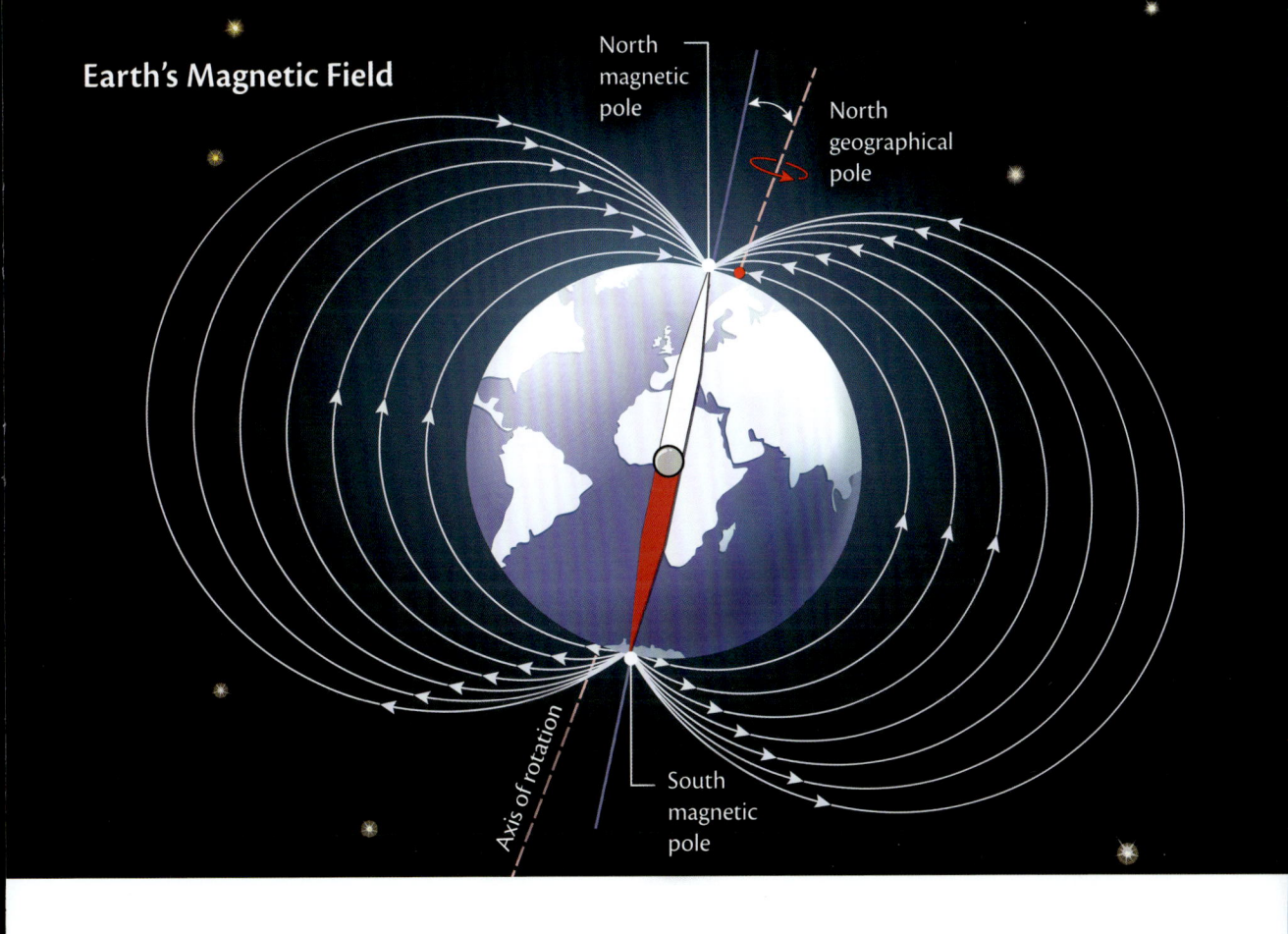

- North magnetic pole
- North geographical pole
- Axis of rotation
- South magnetic pole

down around the South Pole. That is why a magnetized compass needle will point towards the North Pole up in the Arctic. The compass needle tip is magnetic north. Remember, opposites attract. The magnetic north of one magnet will be attracted to the magnetic south of the other magnet. That is why the compass needle always points towards the North Pole.

Actually the location of the magnetic pole up in the Arctic moves all the time—from 6 to 32 miles a year. In the year 2020, the magnetic pole was located about 500 miles (800 km) from the North Pole in the Canadian Arctic.

Nobody really knows what magnetizes the earth like this. It's another mystery designed into our world by the Creator. This magnetic core has helped man find his way around using a compass. In 2004, scientists learned that pigeons can tell where they are by an internal compass. Somehow, these birds can tell the location of earth's magnetic pole wherever they fly.

GOD MADE THE WORLD

When the scientists attached magnets to their beaks, this confused the pigeons and they could not find their way home.

Electrical Energy

"For as the lightening comes from the east and flashes to the west, so also will the coming of the Son of Man be." (Matthew 24:27)

Man has used God's energy from the beginning of the world. He has burned wood to create heat. He has used animals to pull plows. He employed wind energy to push ships around the world. He built water wheels to grind grain into flour. He burned coal to make steam to propel ships and trains.

The trouble with these energy forms is that they were not always dependable. The wind would not always blow, and the water would only flow along certain creeks and rivers. Animals wouldn't always

Lightning Strike in El Paso, Texas

CHAPTER 5: GOD GIVES US ENERGY

cooperate, and they required a lot of work to feed and care for them. Burning wood and coal was messy and sometimes inconvenient. You couldn't store up the energy very well either. Once you started burning the wood and coal, you had to use it right away for heat or mechanical jobs.

To make life easier for all of us, God gave electricity—another use of the electromagnetic force. This electrical energy is easy to store, and it can be harnessed to do all kinds of work for us. It can be used in every home, in every remote place around the world. It isn't messy like burning wood in your furnace, and you don't need to feed it. You don't need to clean up after it either, as you would for oxen and horses. There isn't any weight to electricity, and it can be transported everywhere through wires. Electrical energy can even be stored in batteries, and used in phones and electronic gadgets carried about in purses and pockets.

Electrical energy is the flow of electrons from place to place. Man's discovery of electricity came about when he started studying lightning. God made **lightning** as a very powerful and public demonstration of his power. Basically, this cosmic fireworks display comes when a huge amount of electrons flow from the clouds into the ground. An extra powerful bolt of lightning carrying as much as 100 billion joules that shoots into the ground all at once is called a **superbolt**. From Matthew 24:27, we learn that the coming of Christ in judgment will look like a sheet of lightning that covers the whole world at the same time. Lightning is a symbol of God's power to judge the world.

Benjamin Franklin
(1706-1790)

An early American inventor named Benjamin Franklin made some of very important strides in the discovery of electric energy. Franklin collected electricity from lightning bolts, and coined the terms **battery**, **charge**, **positive**, and **negative**, for the study of electricity.

Although Franklin was not an orthodox Christian, he was also not a Deist. At one point, he boiled down his beliefs to this: "Here is my Creed. I believe in one God, the Creator of the Universe. That He governs it by His Providence. That He ought to be worshipped." 4

How to Make Electricity

There are two kinds of electricity: static electricity and current electricity. The secret to electrical energy is to get the electrons in the atom moving. This will happen if you can separate the electrons from their atom. **Current electricity** is like a kinetic energy because the electrons are flowing through a wire.

One of the uses of electrical energy is the lightbulb, developed by the American inventor, Thomas Edison. The flowing of electrons through a wire heats up a filament, which lights up a bulb.

When the electrons travel through a coil of wire, a magnetic field is created. Since the coil runs through a permanent magnet, the magnetic field created by the coil begins to spin. This is an example of how moving electrons (or electrical energy) can be translated easily into mechanical energy—where objects are in motion and able to do work for us. When you translate electrical energy into mechanical energy or heat energy you can get things done. You can provide mankind with helpful energy. Electricity is a

CHAPTER 5: GOD GIVES US ENERGY

great way to transport energy. And, it can be changed quickly into useful forms. It has made modern life more convenient.

Rub a balloon on your sweater and you can create **static electricity**. By rubbing the balloon, you can actually peel off electrons from the atoms on the surface of your sweater. This makes the surface of your sweater positively charged, while the balloon is negatively charged. So, now the balloon will stick to your sweater. Rub two balloons on your sweater, and they will both become negatively charged. Like charges repel. So a negatively-charged balloon will repel another negatively-charged balloon.

In power plants, electron flow is started up when large copper coils spin inside gigantic magnets. This pulls the electrons away from the atoms in the copper wire and gets them flowing through the wire.

How Much Electricity Would it Take to Kill Somebody?

"When you build a new house, then you shall make a parapet for your roof, that you may not bring guilt of bloodshed on your household if anyone falls from it." (Deuteronomy 22:8)

This Scripture instructs us to fence off dangerous areas where people are likely to congregate. So we should be aware of the dangers of electricity. Parents are careful to warn their little ones to keep fingers out of the electric sockets. When the children are young, dads and moms will put protective plastic plugs into empty wall sockets that aren't being used. Electricity can be dangerous. The Lord wants us to take special precautions when there are dangerous risks in our homes.

The power of electricity is measured by both voltage (volts) and current (amps). Let us picture electricity like a flow of water in a pipe. The quantity of electrons

Balloon Sticks to Wall—Example of Static Electricity

GOD MADE THE WORLD

Electric Pylons

flowing is similar to the volume of water. This is called **current**. The force of the flowing electrons is called **voltage**. This is like the pressure of water in a pipe.

Most household voltage runs between 110 V and 220 V. Is it the voltage (the pressure or force of electricity) that kills? Or is it the amount of current (volume) that kills? The answer is "Both!"

Professionals say that anything over 30 Volts is dangerous. You might have a large pipe with very little pressure pushing the water through the pipe. Or you might have a small pipe with a high pressure of water. Both the amount of water and the amount of pressure pushing the water determines the total energy of it. If you turned a firehose up to full pressure, it would knock you down pretty fast. So, if you have high voltage and a high current (lots of electricity) go through your body, it could knock you down. It might even kill you.

Similarly, you need a certain minimum force from a gun if the bullet will break the skin. Most Nerf ™ guns aren't going to do any damage, because the force of the shot is so low. If the force is strong enough, now the size of the bullet really does make a difference. Even a fairly small bullet could do some damage. One milliamp of current will only provide a little tingle. Between 50 mAmps and 150 mAmps would cause serious pain and possible death. One Amp of flow with anything over 110 Volts could very well result in death.

Controlling Electrical Energy

Electrical energy can be stored in **capacitors**. You can picture a capacitor like a dam holding back water. They are made of two plates separated by an insulator. The negative charges collect on one plate, and the positive charges on the other.

Batteries can also store energy by changing electrical energy into chemical energy. When the battery is ready to be used, it will convert the chemical energy back into electricity.

In order for electricity to start flowing and doing work, you have to complete a **circuit**. To get a circuit, you need three components.

- A power supply of voltage
- A load designed for the power supply
- Wiring that connects it all together in one loop.

If you don't want the circuit running constantly, you will need to add an on-off switch. Water flowing through pipes is controlled by a valve or a faucet. Similarly, the flow of electricity goes through wires, and it is controlled by the switch. When the circuit is open, the electricity cannot flow. Before the switch is turned on, the electrons are waiting like little workhorses pushing against a gate. Then, as the switch is turned on, the electrons begin to rush around the circuit at about 1,860 miles per second! An electrical circuit usually includes a **load**, which would be a large resistor, a light bulb, or a motor. When you plug a lamp into a wall outlet, you provide a load. Now the electrons have a job to do.

Child Safety Cover on Outlet

GOD MADE THE WORLD

They rush around the circuit and heat up the bulb filament, which turns the light on. The circuit is open until you plug the lamp into the outlet (or flip the switch). A **resistor** is a load you add to the circuit to use up excess electricity. It just heats up and reduces the current moving around the circuit.

If somebody puts a screwdriver or a metal wire between the two ends of an open circuit, a **short circuit** (or a **short**) will result. You might see some sparks. It is very dangerous, and it could light the house on fire. Electricity is always looking for a short cut, and it wants to avoid taking on the load if at all possible. The electrons will flow where they can flow the fastest. So if a short circuit develops, the electrons will go for it—lots of them. Because there is no work for the electricity to do in a short, the wires carrying all the electrons will get real hot.

If you are getting shorts in the electric circuits around your house, you will see blackened wires, charred plastic, and burned wood. You might hear small explosions and smell burning wires.

Circuit breakers are supposed to protect your house from these electrical explosions that could burn down your house. Here is how they work. When you get a short in a circuit, a bunch of electrons pour uncontrollably through the short. The current draw of electrons is very high. Every circuit in your house has to go through a circuit breaker. These circuit

Electrical Shorts Cause Fires

CHAPTER 5: GOD GIVES US ENERGY

Circuit Breaker Box

breakers are designed to break the circuit if the current goes too high. They are usually rated for 10 amps, 20 amps, or 50 amps. So when a circuit that is limited to 20 amps develops a short, it will try to draw 100-1000 amps all at once. At that point, the circuit breaker will turn off, because it limits the current flow to 20 amps.

You should be aware of the amount of amps drawn by certain electrical appliances and gadgets in your home. If a circuit is limited to 20 amps, you cannot use a bunch of electrical devices which consume more than 20 amps on the same circuit. The following is a table that shows the typical electrical current used for common electrical devices. How many kitchen devices could you use on one 20 amp circuit?

Typical Amperage for Household Devices	
Clothes Dryer (Electric)	20-30 Amps
Toaster	12-14 Amps
Washing Machine	12 Amps
Clothes Iron	10 Amps
Portable Heater	10 Amps
Microwave Oven	10 Amps
Dishwasher	10 Amps
Vacuum Cleaner	10 Amps
Hair Dryer	10 Amps
Coffee Maker	9 Amps
Computer	4-7 Amps
Blender	4 Amps
Can Opener	1 Amp
Lamp	0.25 Amps

How to Be Safe With Electricity

Remember that there are materials that conduct heat, and there are materials that insulate and do not conduct heat. Similarly, there are materials that conduct electricity which are called **conductors**. There are also materials that do not conduct electricity. These are called **insulators**.

The best material for conducting electricity is silver, but it is expensive. Silver is 5% better at conducting electricity than copper. Gold and aluminum are good conductors, but not nearly as good as copper.

Good insulators are made of material where the atoms hold on to their electrons pretty tightly. That is, the electrons can't be jarred out of orbit very easily. The best insulators are made of glass and porcelain (made of clay and quartz).

High voltage electric wires need really good insulators, because you don't want people touching the wires. When humans touch electric wires, their bodies can become a conductor, passing electricity through their bodies to the earth. Electricians usually wear boots with rubber soles to protect them from electric shock. Silicone rubber is the best insulator for screwdrivers and gloves used by electricians.

Energy Costs

The lazy man does not roast what he took in hunting, but diligence is man's precious possession. (Proverbs 12:27)

Every home uses electricity these days, and this energy costs money. Most electricity plants burn coal or oil to make electricity. Some electric plants (Hydroelectric Plants) run their turbines on water running down through dams. While expensive to build, these plants do not need to buy fuel to run them. It also costs money to maintain electric plants, and distribute the electricity to homes.

Electricity Insulators

CHAPTER 5: GOD GIVES US ENERGY

Most families spend between $50 and $500 per month on electricity. We should be careful not to waste the resources that God gives to us. Proverbs 12:27 points out that slothful men tend to be wasteful. We are called to be good stewards of the resources God gives us.

The following table shows where most of the money for energy goes in our homes. This example assumes that a family spends $300 per month (average) on energy.

Electric Bill

Household Energy Use	Percent of Total Use	Average Monthly Cost
Heating and Air Conditioning	46%	$138
Water Heating	14%	$42
Refrigerator, Washer, Dryer	13%	$39
Lights	9%	$27
TV, Computers, and Media	4%	$12

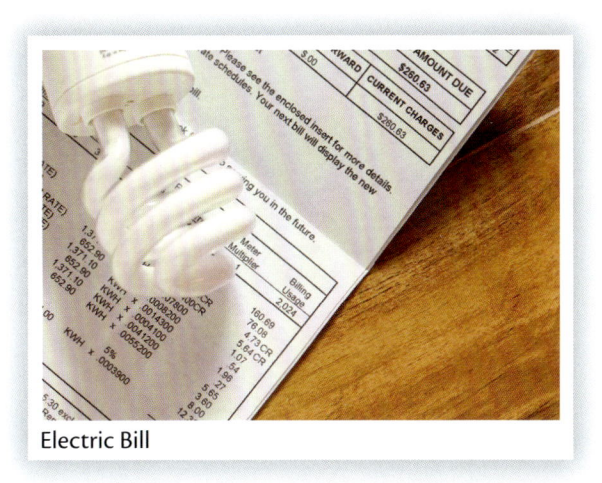

Saving Energy in Heating and Cooling Your House

Heating and cooling is always the biggest energy cost in the home. An air conditioner can use up 1,000 - 2,000 kilowatt-hours per month. At 20 cents a kilowatt-hour of energy, this would cost $200 - $400 per month. Here are some good ways that every family member can help conserve energy and save money in heating and cooling the home.

1. Be sure there is good insulation in the ceiling and walls, especially the outside walls.
2. Open and close the front door quickly when you go in and out. A lot of heat escapes the house if you leave the door open if the house is heated. During the

Digital Thermostat

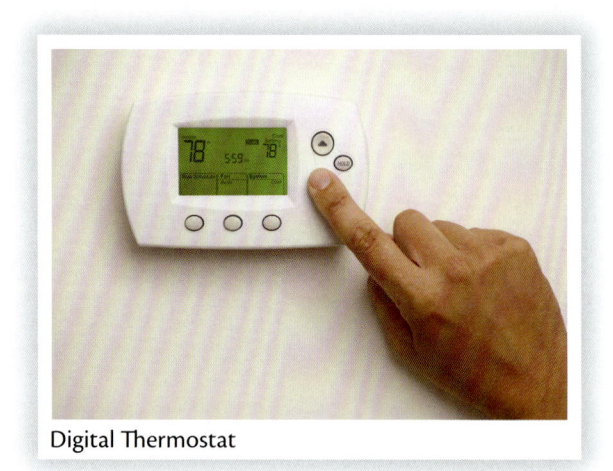

summertime, heat from the outside will quickly invade the cool air inside through an open door.

3. Remember, heat rises. If you have ceiling fans, be sure they are pushing cool air downward during the summertime by running them counterclockwise (as you look up at them.) In the wintertime, run the ceiling fan clockwise to pull the cool air upwards, and keep the warm air pushing down.

4. Keep the thermostat at about 78°F in the summer and keep it under 65°F in the winter. Wear warmer clothes in the wintertime.

5. Keep the window coverings in the sun-facing windows closed in the summer and open in the winter.

6. Regularly replace air filters in your heating and air conditioning systems.

7. Don't block vents that are taking air into your heating or air conditioning systems.

8. Do an inspection of windows, doors, and attic entries. Identify all locations where hot or cold air is slipping into the house, and seal with weatherproofing.

9. About 30% of a house's hot or cool air disappears through single-paned windows. Consider replacing with double-paned windows, which provide an insulating air gap between the two panes. This could provide a savings as high as $100 per window per year.

10. An efficient wood stove or fireplace can reduce money spent on gas by as much as $200 per month. This is especially true if your family has access to a free source of wood to burn.

Saving Energy with Your Water Heater

The water heater is another big energy user in your home—second only to the heating and cooling system. At 405 kilowatt-hours a month, water heaters

CHAPTER 5: GOD GIVES US ENERGY

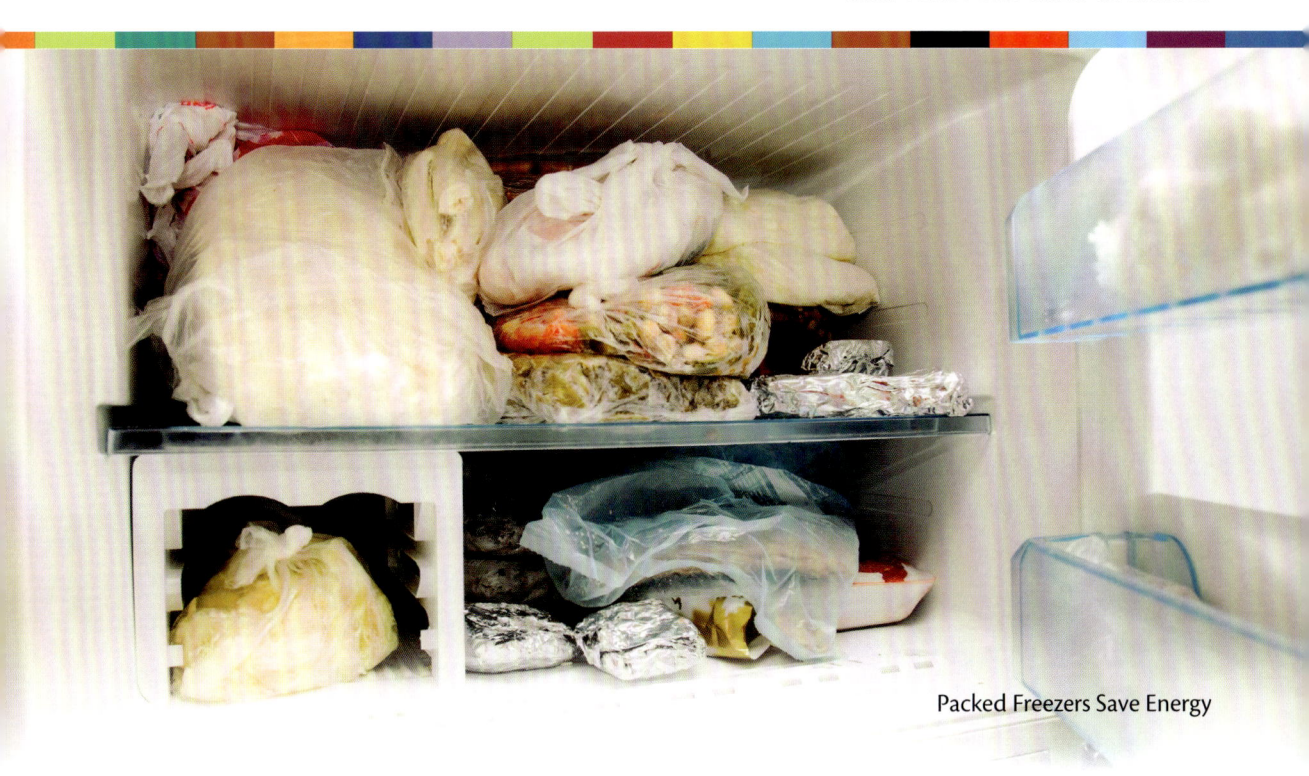

Packed Freezers Save Energy

can cost about $80 per month to operate. Here are some tips for keeping the water heater running efficiently.

1. Take shorter showers. Although you may enjoy the luxury of a long, warm shower, remember that showers cost money. A five-minute shower consumes about ten gallons of hot water, whereas a twenty-minute shower uses about 40 gallons.

2. Make sure that the water heater is covered with an insulation jacket. Also insulate the hot water pipes that are visible in the utility room.

3. Turn down the water heater when you go on vacation. Also, you can put the water heater on a timer, only keeping it on when people are likely to take showers.

4. Set the water heater's temperature to 120°F or lower.

5. Install shower heads and faucet aerators that conserve water.

Saving Energy with Appliances, Lighting, and Electronics

Although appliances, lights, and electronics do not take as much energy, there

171

may be easy ways to cut costs here as well. Families must be careful, though. Modern appliances and energy sources are meant to make life more convenient. Some people do spend too much time trying to save money, and they end up making their lives more inefficient. However, there are some money-saving tips that are not so troublesome, and can save money.

1. Keep your freezer pretty well stuffed and your refrigerator not quite filled up all the way. Frozen goods help keep the freezer cold, but overstuffing the fridge actually increases energy usage.
2. Don't leave the fridge door open very long. Also, keeping things organized means less time looking for food.
3. Consider using cold water for doing the laundry.
4. Turn off lights when they are not being used. Consider getting a universal switch that turns off all lights and non-essential energy-consuming devices when you leave the home.

The Power of God

After these things I heard a loud voice of a great multitude in heaven, saying, "Alleluia! Salvation and glory and honor and power belong to the Lord our God!" (Revelation 19:1)

In all of this study of energy and technology, let us still remember that *we don't create energy*. We are harnessing God's energy. Years ago, a farmer would harness his oxen, and they would pull his plow. These were God's oxen eating God's grass, and producing God's energy. The farmer was only harnessing the energy God gave him, and he was putting it to good use.

That is what we do with electricity, water power, wind power, fuel, and heat. We have learned to store energy and use it where and when we want it. And, we are only using a tiny part of the energy of

Water Heater Temperature Control

CHAPTER 5: GOD GIVES US ENERGY

the universe. How could the universe ever run out of energy? The stars are built to burn for a trillion years. And, God could make more stars if He wanted to.

We are thankful for the power plants that can turn on lights for a whole city, keep everybody warm in the winter, and keep the food cold in the refrigerators. We are thankful for powerful engines that can take a car from 0 to 70 mph in 5 seconds. But, the earth and Jupiter are much, much bigger than a car. How did that much mass begin to travel at 100,000 miles per hour through space? Never forget, it was God that did it. Let us praise Him for His limitless power! ■

GOD MADE THE WORLD

Pray

- Praise God for His great power. Praise Him for lightning and for electrical power. Praise Him for His scientific laws. Give Him the glory for how everything works together. Praise Him for the forces of the universe that keep atoms together and planets spinning and revolving. Praise Him that even the most intelligent scientists are baffled by the complexity of the universe!
- Thank the Lord for His faithful servants, scientists to whom He has given insight, and scientists who give Him the glory.
- Thank Him for the energy that we use every day for heat, for lights, for operating our cars, and for washers, dryers, and fridges.

Sing

My God is So Big, So Strong and So Mighty

My God is so big, so strong and so mighty
There's nothing my God cannot do
My God is so big, so strong and so mighty
There's nothing my God cannot do.

He made the trees
He made the seas
He made the elephants too

My God is so big, so strong and so mighty
There's nothing my God cannot do
My God is so great, so strong and so mighty
There's nothing my God cannot do
My God is so great, so strong and so mighty
There's nothing my God cannot do.

The mountains are His
The rivers are His
The skies are His handy works too

My God is so great, so strong and so mighty
There's nothing my God cannot do
There's nothing my God cannot do
There's nothing my God cannot do
For you.

If you do not know the hymn, you may listen to a version of the hymn on the Internet, with supervision, and sing along with it.

Do

Choose at least one of the following activities and apply the lessons you have learned in this chapter.

1. **Survey your house for the condition of the insulation.** Check out the weather-stripping on the doors and windows. Using an accurate thermometer, measure the temperature on the inside of each window. Feel to see if there is a breeze, cool air, or warm air entering the house through doorways and windows. Investigate ways you can improve the insulation of your home economically. If your improved plan could save the family $20.00 per month on energy, would it be worth it to do your plan? Consider also new and improved windows. How much could you save in energy costs if you improved the windows in your home?
2. **Check for electrical safety in your home, with parental supervision.** Are all the electrical outlets within reach of two-year-old children covered with plastic plugs? Do you see any evidence of shorts? Is there any exposed electrical wire (where the copper is showing) around the house?
3. **Review the Circuit Breakers in your house, with parental supervision.** What is the rating in each room or area in the home? What are the electrical devices and appliances you use most often? How many Amps does each take? In what rooms would you be most likely to trip a circuit breaker?

Watch

To watch the recommended videos for this chapter, go to **generations.org/ GodMadeTheWorld** and scroll down until you find the video links for Chapter 5. Our editors have been careful to avoid films with references to evolution. However, we would still encourage parents or teachers to provide oversight for all internet usage. These videos may not give God the glory for His amazing creative work, so the student and parent/teacher should respond to these insights with prayer and praise.

Car Crossing the Bixby Canyon Bridge, Big Sur, California

Chapter 6
GOD'S DESIGN OF MOTION

> Oh, that You would rend the heavens! that You would come down! That the mountains might shake at Your presence—as fire burns brushwood, as fire causes water to boil—to make Your name known to Your adversaries, that the nations may tremble at Your presence! When You did awesome things for which we did not look, You came down, the mountains shook at Your presence. For since the beginning of the world men have not heard nor perceived by the ear, nor has the eye seen any God besides You, who acts for the one who waits for Him. (Isaiah 64:1-4)

The true God, our Creator, is the living God. He is a God of action. He works. He does powerful things. And, we see His works all around us. But He also gave His creation the ability to move, to act, and to work. He made things so that we could move over land, water, and air. He created us with the ability to walk, run, and jump. Various kinds of His animal creation can run, swim in the water, and fly through the air.

By His generous mercies, God gives us things to help us move around. Horses, donkeys, elephants, mules, and even dogs have helped man to go places by carrying people on their backs or pulling wagons.

There are different ways to go places, and to move around God's world.

We use our own muscles. Sometimes we will use our muscles, but we are helped

GOD MADE THE WORLD

by a **mechanical advantage**. Man figured out ways to use mechanical inventions to move himself or to move other things. Examples of this are levers and pulleys used for rowing a boat, winching a heavy car out of a ditch, or riding a bicycle.

1. We use the muscles of animals when we ride horses or use Siberian huskies to pull a dogsled.

2. We use a motorized vehicle that relies on another energy source such as gasoline or electricity.

Mechanical Advantage

Thankfully, God gave humans brains to invent machines. These inventions help to make life easier for us. The **wheel** is one of the best examples of a simple machine.

What if you tried to drag a 100-pound (45 kg) friend along the ground for 30 feet (9 m)? Would it be easier if you put your friend in a wagon and then pulled it the same distance? The wheels on the wagon are what makes the work easier for you. In fact, it would take about 40 times more energy and strength for you to drag your friend than it would to pull him in

CHAPTER 6: GOD'S DESIGN OF MOTION

a wagon. The ground is causing more friction against the body of your friend, much more than the resistance against the four tires on the wagon.

Do you think you could lift your 100-pound (45 kg) friend six feet (1.8 m) in the air? Yet, if your friend was sitting on a teeter-totter, you could push or pull down on the other side and lift him up. This is called **leverage**. This is another example of mechanical advantage.

The Bicycle

The bicycle is a great blessing from the Lord. To this day it is a preferred form of transportation for humans.

Vehicle	Number in the World
Bicycles	2 billion
Cars	1.4 billion
Motorcycles	0.25 billion
Horses	0.06 billion

The greatest blessing of the bicycle is its efficiency. This means that the bicycle doesn't waste very much energy. You will use five times more energy walking than riding your bike. Cars are even less efficient. You could ride a bike for 57 miles on the energy it would take to drive a car for one mile. While a human can travel 100-200 miles a day by bicycle, he would only make it 20-25 miles by walking.

The fastest bicycle ride (downhill) on record happened in 2017, when Eric Berone road his bike at 142 miles per hour (228 kph). That's quite a bit faster than

Mountain Biking Down a Mountain

Usain Bolt's top running speed of 28 mph.

So what makes the bicycle so efficient? The easy answer is that it doesn't waste energy when compared to walking. Part of the energy for walking is used by your legs holding up the rest of your body. Also, have you noticed that when you walk you will lift your feet off the ground? Energy is wasted when you have to fight gravity for each step you take. When the bicycle rolls over the ground, there are only two small spots at which the bicycle is touching the ground. While the bicycle must overcome this friction force, a small amount of pressure on the pedals helps the rider move the bicycle forward. Also, once you get a bicycle going, you can coast for awhile without applying additional energy. The bicycle is a very useful machine to help humans get around.

Moving Across the Land

What does it take to move something across the floor or down a road? You have to apply a **force** to get it moving. Force is what it takes to get something moving or

CHAPTER 6: GOD'S DESIGN OF MOTION

to get it to move faster. Before we think about force, we need to look at movement. When you want to move something, you have to speed it up from 0 mph to a certain speed. When something is speeding in a certain direction, the measurement of the speed is called **velocity**. When you see the letter "v," in physics, that usually means velocity. If the velocity is 0, then the car or the bicycle is standing still.

To speed a car or a bike up from 0 to 10 mph (16 kph), you would have to accelerate or "speed up." You have to increase the velocity by accelerating. **Acceleration** is usually referred to by the letter "a" in physics.

When something is speeding up, it is in **positive acceleration**.

When something is slowing down, it is in **negative acceleration** or **deceleration**.

Now, how do you get a car to speed up from 0 to 10 mph, or from 10 mph to 20 mph? You have to apply a force. This is the only way to do it. If there is no force applied, there will be no acceleration.

It also takes force to overcome the friction on the tires of a car or bicycle. The force must come from the car engine, or from the bicyclist's legs. The friction is pushing against the tires. If the tires are sunk into mud, the friction is going to be stronger. It will take more force to push against the friction of the mud, to break the wheels loose and get the vehicle moving.

If the car is starting up a hill, the engine will have to work harder than it would if the car was traveling over level ground. The engine would have to push against

It Takes Force to Get a Car Going and to Push a Car Uphill

GOD MADE THE WORLD

the gravitational force that is trying to get the car to roll back down the hill. If the car is sunk in the mud and going up hill, the engine will have to work even harder to roll the car forward.

If you are riding a bicycle through the sand, it takes a lot of hard work to keep it going. That's because the friction, or resistance, is very high when peddling through sandy soil. If you are riding your bike on smooth pavement, it doesn't take as much work. The peddling force to overcome the frictional force would be much less.

To get work done, you will always need to apply a force. And it takes energy sources to get the forces doing the work. Force is measured in Pounds-Force (in the English measurement system), and Newtons (in the Metric system).

Isaac Newton's **Second Law of Motion** explains the relation of movement and force. Mr. Newton figured out that it takes a constant force on a body to create a constant acceleration, or to increase its velocity. This is the formula that comes from this law of motion:

Bicycling Through Mud and Sand Requires More Force (and Work)

Bugatti Veyron—Fastest Street-Legal Production Car in the World
1000 Horsepower and Capable of 268 mph

$$F = ma$$
or
Force = mass x acceleration

Let's say you want to accelerate a car that weighs 5,000 pounds, so that it will speed up to 70 mph in just 5 seconds. The acceleration would have to be about 23.5 ft/s^2. How much force would it take to speed a car up to 70 miles per hour in five seconds?

Using Newton's formula, we multiply 5,000 times 23.5, and we get 117,500 Pound-Force (or 523,000 Newtons).

(A Newton is a kilogram—meter/sec^2, whereas a Pound-Force is a pound—foot/sec^2. To make this example easier, we did not calculate the force it would take to overcome the air resistance or the friction on the tires.)

Americans like to measure force in horsepower, because they can picture the amount of force a little better that way. To give you an idea of the size of the force for accelerating the car, 117,500 pound-force is equal to 214 horsepower. Actually, one horse can apply the force of about 15 horsepower. That means it would take the force of 14 horses to speed a car up to 70 miles per hour in just five seconds. Most cars would take about ten seconds to get to 70 mph.

The fastest car in production in 2016 was the Tesla Model S P100D. The engine in this car produces about 500 hp of force, and it can get to 60 mph in 2.5 seconds. There's a very powerful force in that engine! The Bugatti Chiron broke a record in 2017 by accelerating from 0 - 250 mph (400 kph) in 42 seconds. The engine in this car produces about 1,500 horsepower. Think of 100 horses stuffed into that engine working together to provide that amount of power!

What is the fastest thing in the universe? A lot of things run very fast in God's incredibly powerful universe. For example, Mercury is the fastest running planet, speeding around the sun at 112,000 miles per hour (180,000 kph). That's not the fastest object in the universe. There's a whole galaxy called M87 that is shooting towards us at a speed of 638 miles (1,026 kilometers) per second (or 2,300,000 mph).

Let us never forget that it was God who made all these things to go fast. He gave us energy. He created our muscles, powerful horses, and fuel to burn. These were the Lord's words to Job, as He spoke of the power and speed of the battle horse:

Gravity Causes Leaves to Fall to Ground

"Have you given the horse strength?
Have you clothed his neck with thunder?
Can you frighten him like a locust?
His majestic snorting strikes terror.
He paws in the valley, and rejoices

CHAPTER 6: GOD'S DESIGN OF MOTION

in his strength;
He gallops into the clash of arms.
He mocks at fear, and is not frightened;
Nor does he turn back from the sword.
The quiver rattles against him,
The glittering spear and javelin.
He devours the distance with fierceness and rage;
Nor does he come to a halt because the trumpet has sounded."
(Job 39:19-24)

Falling Objects—Gravitational Force

As already mentioned earlier, God made the gravitational force when He made everything. Although we live with gravity every day, and we are very familiar with it, it is still a mysterious force that baffles scientists.

People fall. Things fall off the table onto the floor. Birds fall out of the sky. The force of gravity causes things to fall.

"Are not two sparrows sold for a copper coin? And not one of them falls to the ground apart from your Father's will." (Matthew 10:29)

Even though God has ordained certain natural laws, He is still always in control. Nothing will fall to the ground unless God causes it to happen.

When you drop a ball to the ground, it starts out at 0 mph. Then, it begins to fall faster and faster. That is because the ball is accelerating (speeding up). Cars accelerate on the road, and balls accelerate when they are dropped. When you are on the earth, the earth's gravity provides the force and the acceleration for things that are dropped. The measurement for this acceleration is pretty constant around the world—32 feet/second^2 or 9.8 meters/second^2.

If you dropped a ball on the moon, it would drift to the ground slowly. The moon has less gravity pulling on the ball because the moon is much smaller than the earth. If you dropped a ball on the sun, it would fall about 30 times faster than it would on the earth.

GOD MADE THE WORLD

Gravitational Acceleration on Some Celestial Bodies	
Sun	274 m/s²
Moon	1.62 m/s²
Mars	3.77 m/s²
Jupiter	26 m/s²

If you dropped a bowling ball on your foot here on earth from about four feet in the air, you could very well break a bone. This really does happen to bowlers sometimes. Now, if you dropped a golf ball on your foot on the sun from about four feet high, you would do about the same amount of damage to your foot. It would probably break a bone. The bowling ball would be falling about 8 feet/second when it hit your foot on earth. The golf ball would be falling at about 80 feet/second when it hit your foot on the sun. The increased speed of that golf ball on the sun comes by the increased force of gravity on the sun.

Work

The scientific concept of **work** is a simple idea. It is the application of force over a distance. If you push a piano one inch across the floor, you haven't done very much work. But, if you push a piano for a mile, you have done a lot of work. The greater the distance you apply the force, the more work is getting done. Applying a little force to a little marble over a short distance hardly takes any work at all. Work can be expressed in a mathematical formula like this:

Work = Force x Distance

Friction is Helpful

Friction is not all bad. In fact, we should be thankful for friction. The Lord created friction for a lot of helpful purposes. For example, it's really hard to

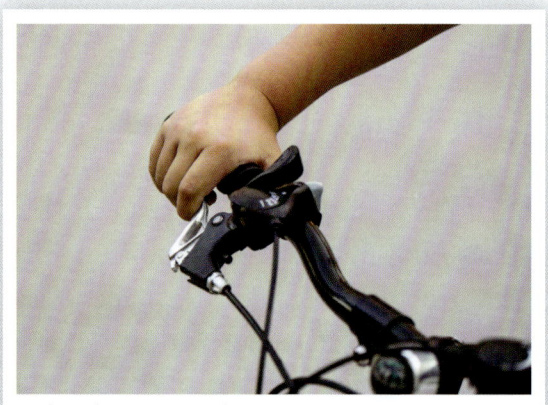

Brakes Slow Down Bikes Using Friction

CHAPTER 6: GOD'S DESIGN OF MOTION

ride a bicycle, or drive a car on ice. That's because the harder you peddle your bike, the more your wheels slip on the ice and you go nowhere while your wheels are spinning.

Also, brakes use friction to slow the wheels down. Examine your front brakes on your bike. You can see that the brake pads rub against the wheel when you pull the brake. This is another good use of friction.

Opening a jar lid requires friction between your hand and the lid. If your hand was coated in slippery oil or peanut butter, you could not open the can very well. God created your fingers with

Front Wheel of a Bicycle

ridges called finger prints that provide a roughness, which gives you more friction when you open can lids. Also, pencils use friction to make marks on paper. The

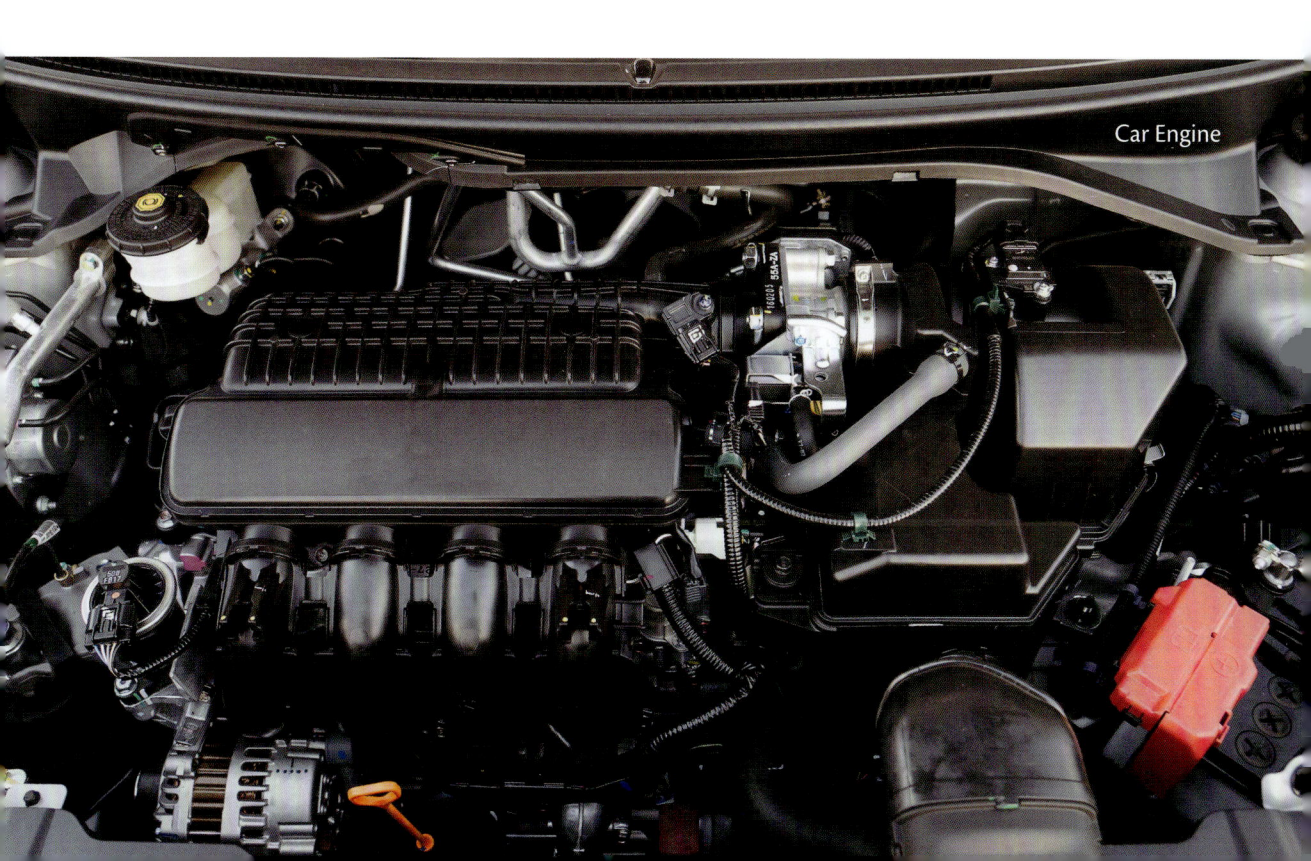

Car Engine

Combustion Engine Diagram

Fuel → 1. Intake
2. Compression
3. Power
4. Exhaust → *Exhaust*

graphite rubs off onto the paper using friction.

Making Cars and Bikes Run Smoothly

While friction can be helpful, sometimes it can be hurtful too. The wonderful wheel can roll over rough surfaces pretty well, which overcomes the friction that would prevent its moving. The wheel spins on an axle or a rod that slides through the center of the wheel. You'll notice that your front bike wheel spins fast and freely most of the time. However, sometimes there is unwanted friction between the axle and the wheel, making the bike run inefficiently. This could be the brakes rubbing against the wheel, and sometimes it is the axle rubbing against the wheel hub. One of the best ways to keep your wheels spinning freely is to use **lubrication**. When the axle rubs against the wheel hub, that slows the wheel down, and it wears out the axle. Lubrication makes the contact between the wheel and the axle slippery. Keep your bike running smoothly by using axle grease for the wheel axle.

Most wheels spin on ball bearings or roller bearings which are placed between the axle and the wheel. Scientists found that ball bearings can be one of the best ways to reduce friction between the axle and the wheel.

CHAPTER 6: GOD'S DESIGN OF MOTION

Engines—Powering Vehicles for Land

Bicycles are powered by human muscles, but cars, motorcycles, trucks, buses, and trains are powered by motors and engines. Motors run on electricity, and engines run on gasoline or diesel fuel.

Most cars are powered by an engine using gasoline, but carmakers are making more cars now powered by electric motors. At this time, about 4% of cars in the world are electric. That's about 30 times what it was ten years ago.

Most cars and trucks use an "internal combustion engine." This kind of engine burns fuel in a cylinder, which increases the pressure in the cylinder. This causes pistons to move up and down, which moves a crankshaft, which moves the axles, and that is what turns the car's wheels.

Engines used for small cars will either use four or six cylinders. Big SUV's and trucks require bigger engines, where they will use eight or ten cylinders. In most internal combustion engines, the pistons go up and down in the cylinder using four strokes:

1. The intake stroke takes gas and air into the cylinder through a valve, as the piston goes down.

2. The compression stroke closes the valves. Then, the piston goes up and puts high pressure on the gas and air.

3. The power stroke happens right after the spark plug lights the fuel, causing a burst of heat and an increase of pressure in the cylinder. The pistons slam down, while the valves are still closed.

4. The exhaust stroke happens as the piston comes back up. The exhaust valve opens, and the extra gases escape.

Newton's Pendulum—For Every Action There is a Reaction

GOD MADE THE WORLD

These gases pass through the muffler out the exhaust pipe.

Both diesel and gasoline-powered engines used in cars and trucks are internal combustion engines. The main difference between them is that the diesel engine does not need spark plugs. Instead, it is the pressure on the compression stroke that ignites the fuel mix.

Moving Across Water

And God said to Noah, "The end of all flesh has come before Me, for the earth is filled with violence through them; and behold, I will destroy them with the earth. Make yourself an ark of gopherwood; make rooms in the ark, and cover it inside and outside with pitch." (Genesis 6:13-14)

God made water, and He made fish with the ability to swim in that water. The Creator also intended for man to make boats, like the ark which saved Noah and his family in the great flood. God gave men the wisdom to figure out how to propel their boats through the water.

Isaac Newton came up with the **Third Law of Motion** which explains the way in which fish swim and boats are propelled. The law is stated this way:

To every action there is an equal and opposite reaction.

Orville Wright (1871-1948)

Wilbur Wright (1867-1912)

First Flight at Kitty Hawk, 1903

When somebody fires a gun, there is the movement of a bullet out of the barrel of the gun. But the gun also slams back in the opposite direction. This is called **recoil**. It is an example of **Newton's Third Law**.

Swimming gives another good example of this law of motion. When you apply the force of your hand against the water, your body is pushed forward in the water. The **action** happens when the water is pushed backward. And, the **reaction** happens when your body is pushed forward.

Rowing a boat works much like swimming. The action of the oars moves the water backward. The reaction occurs when the boat moves forward.

Most boats today are pushed along in the water by gas engines. Boat engines turn a fan-like **propeller**. The propeller moves through the water like a screw or drill bit moves through wood. The propeller pushes the water backward behind the boat. This is the action. The boat reacts by moving forward.

Moving Through the Air

There are three things which are too wonderful for me,
Yes, four which I do not understand:
The way of an eagle in the air,

GOD MADE THE WORLD

The way of a serpent on a rock,
The way of a ship in the midst of the sea,
And the way of a man with a virgin.
(Proverbs 30:18-19)

God created birds to fly through the air. Truly the genius of flight comes from the Creator. For 6,000 years, nobody could figure out how to do it. Proverbs 30 talks about the very deep mysteries involved with flight and fluid mechanics (the way boats cut through the water). The men who came up with the first flying machine had to study God's design of bird flight for a long time. Orville and Wilbur Wright, sons of a Christian pastor, tried many different experiments with flying, based on their study of birds. They worked on many different designs of the airplane. It would crash on the ground, and they studied birds some more until they had an airplane that could fly. On December 17, 1903, at a place called Kitty Hawk, North Carolina, the Wright brothers successfully flew a craft for 120 feet. On the same day, they tried another flight, and went 852 feet over 59 seconds.

Then God said, "Let the waters abound with an abundance of living creatures, and let birds fly above the earth across the face of the firmament of the heavens." So God created great sea creatures and every living thing that moves, with which the waters abounded, according to their kind, and every winged bird according to its kind. And God saw that it was good. (Genesis 1:20-21)

Flying is a little bit like swimming. When a bird flies, its

Jet Airplane

192

wings push the air backwards. This is the action. The reaction is the bird moving forward in the air.

Suppose a boy on roller skates is given two large pieces of cardboard. He holds one in his left arm and the other in his right. When he pushes the cardboard pieces backward quickly what action is produced? Of course, the air is pushed backward. Then, what is the reaction? His body moves forward on the skates.

In order for an airplane to fly through the air, two things are needed: **thrust** and **lift**. The airplane has to move forward with a thrust. And it has to be able to fly upward by a lift.

In order for a car to move forward it must apply a force by spinning its wheels against the ground. Airplanes cannot push against the ground when they are in the air.

Scientists have come up with two ways to push an airplane forward in thrust. One way is to use a propeller similar to a boat. The propeller pushes the air backward (action), which makes the airplane go forward (reaction).

The second way to get thrust for an airplane is using jet engines. 23-year-old British engineer Frank Whittle invented the jet engine in 1930. The jet engine burns fuel, and the exploding gases rush out the back end (action). The reaction is the jet plane moving forward.

Fuel needs oxygen to burn. In a jet engine, compressor blades push lots of air into a chamber. The fuel pours into the same chamber, mixes with the oxygen, and burns. The pressurized gas rushes through turbine blades, turning the blades which turn the compressor. This keeps up the pressure on the air rushing into the jet engine. Once the gases pass through the compressor they exit the back of the jet.

Diagram of Wing

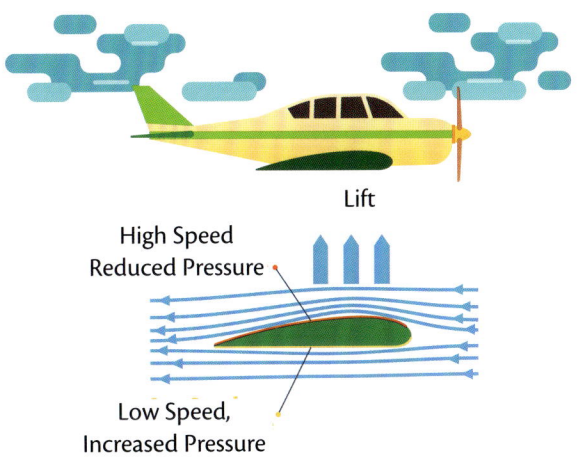

GOD MADE THE WORLD

How Airplanes Get Their Lift

In addition to forward motion, an airplane must be able to lift up into the air for flight. There are two forces that must be overcome—gravity and air pressure.

Air is always pressing in on us from all directions. For an airplane to go up in the air, the air pressure underneath must be greater than the air pressure pressing down from above.

Several simple practical examples will help you to understand lift. Fold a piece of paper in half, hold the bottom half with the flap facing towards you. Pull the paper toward you quickly. What happens to the top flap? Of course, it flies upwards. That is because the pressure under the flap is greater than the pressure on top of the flap.

Place a ribbon in a closed book, allowing about three to six inches to hang down. Hold the book in front of you so the ribbon flops on the side of the book facing away from you. If you blow across the top of the ribbon, it will rise up. This is called the **Bernoulli Principle**, which states that increased speed of the fluid (or air) will lower the pressure of the fluid (or air).

If the air is blowing faster over the top surface of something like a piece of paper, this will create less pressure on that surface. The pressure underneath will be higher than the pressure on the top surface of the paper. Because the pressure is higher underneath, that causes the paper to rise up. So

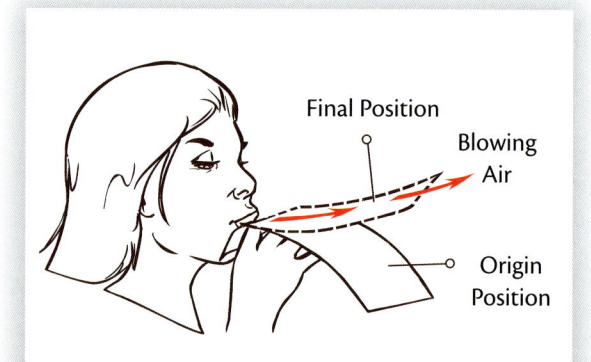

Bernoulli Principle—Illustrated by Blowing Over Top Surface of Paper

194

Space Shuttle

in the example above, if you blow air super fast across the top of the ribbon, you will get less pressure on top of the ribbon. And there will be more air pressure on the under side of the ribbon. If there is more pressure underneath the ribbon than above the ribbon, the pressure under it will push the ribbon up in the air.

When a wing is shaped like the one in the diagram, the air flow above will travel faster than the air flow beneath it. That causes the wing to lift up. That's why God designed the bird's wing this way, and humans have copied this for the wing design on an airplane. Scientists still don't know for sure why God's wing design increases the flow of air on the top side while decreasing the flow on the bottom side.

"For My thoughts are not your thoughts,
Nor are your ways My ways," says the LORD.
"For as the heavens are higher than the earth,
So are My ways higher than your ways,
And My thoughts than your thoughts."
(Isaiah 55:8-9)

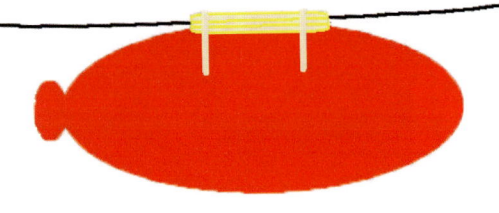

Build Your Own Balloon Rocket Using a String, a Straw, Tape, and a Balloon

Rockets

When man wanted to travel beyond the earth, he had to think of a new way to create motion. Airplanes and jets use the air that surrounds our globe to create the action and reaction. Rockets are made to go into space, where there is no air (and no oxygen). Propellers wouldn't do any good in space, because there would be no air to push through the propellers. There would be no action, and therefore no reaction. So the only way to thrust a rocket forward is to create pressurized gases inside, which blow out the backside of the rocket.

You can create a simple rocket with a balloon. Blow it up, but don't tie it off. Just let it go, and your little rocket will shoot around the room. The pressurized air inside the balloon blows out the back side. That's the action. The balloon rushes forward in the reaction.

The rocket carries fuel and oxygen with it. It is the fuel that burns, and oxygen is needed to burn fuel. It takes a lot of fuel to get someone to the moon. To get the first three men on the moon, the 1969 Apollo 11 mission needed 500,000 gallons of fuel. A family car designed to transport eight people will hold about 30 gallons. So, rocket ships are much larger than cars and trucks. The fuel can weigh twenty times what the orbiting vehicle weighs.

It takes quite a bit of fuel to get a rocket out of the earth's gravitational pull. Once the rocket is in space, nothing can slow it down because there is no air resistance. However, rockets will help the craft to speed up in space. It took the Apollo 11 astronauts about 76 hours to make it into the moon's orbit at an average speed of 3,000 miles per hour (5,500 kph).

Rocket fuel must burn very quickly, without exploding and blowing up the craft. Sometimes rockets use solid fuel, but most rockets today use liquid fuel like gasoline. The gasoline is mixed with liquid oxygen in the combustion chamber. The gases are pushed through smaller nozzles at 5,000 to 10,000 miles per hour, producing the action. The reaction is the rocket launching upward into space.

CHAPTER 6: GOD'S DESIGN OF MOTION

Pride in Technology

Woe to those who go down to Egypt for help,
And rely on horses,
Who trust in chariots because they are many,
And in horsemen because they are very strong,
But who do not look to the Holy One of Israel,
Nor seek the LORD!
Yet He also is wise and will bring disaster,
And will not call back His words,
But will arise against the house of evildoers,
And against the help of those who work iniquity. (Isaiah 31:1-2)

Think of all the changes that have happened in the world over the last one hundred years. Since the year 1900, scientists have introduced rocket ships, jet airplanes, automobile manufacturing, computers, telephones, television, and electrical systems. With all this technology, man has become very proud. Since the 1960s, prayer has been outlawed in most schools around the world. In many places it is against the law to kneel down in the classroom and thank God for His good gifts. Many of the modern schools, businesses, and governments do not fear God and worship Him anymore.

The warning from Isaiah 31 is very real for all of us today. This technology cannot save us. We must seek the Lord, or He will bring a disaster upon our nations. We must worship Him and thank Him always for His good gifts. ∎

Pray

- Let us humble ourselves, and ask God's forgiveness for our pride. We have been too proud of our technology and our own brains. We have not been as grateful and humble as we should be.
- Thank God for the technology of engines, cars, bicycles, motorcycles, buses, airplanes, boats, telephones, and computers.
- Thank Him for the godly scientists who brought us physics, chemistry, and medical discoveries.

Sing

Praise the Lord! Ye Heavens Adore Him

Praise the Lord! ye heav'ns, adore Him;
Praise Him angels, in the height;
Sun and moon, rejoice before Him;
Praise Him, all ye stars of light.
Praise the Lord! for He has spoken;
Worlds His mighty voice obeyed;
Laws which never shall be broken
For their guidance He has made.

Praise the Lord! for He is glorious;
Never shall His promise fail;
God has made His saints victorious;
Sin and death shall not prevail.
Praise the God of our salvation!
Hosts on high His pow'r proclaim;
Heav'n, and earth, and all creation,
Laud and magnify His name.

Worship, honor, glory, blessing,
Lord, we offer unto Thee;
Young and old, Thy praise expressing,
In glad homage bend the knee.
All the saints in heav'n adore Thee,
We would bow before Thy throne;
As Thine angels serve before thee,
So on earth Thy will be done.

If you do not know the hymn, you may listen to a version of the hymn on the Internet, with supervision, and sing along with it.

Do

1. **Lubricate your bicycle's moving parts.** (Or with a little bit of help from your parents, lubricate the bearings on your car.) To lubricate your bicycle bearings, use a lithium complex, soap-based grease. To lubricate your bike chain and gears use bike chain lubricants where bikes are sold. With supervision, you can get more specific instructions on how to do this on the Internet.
2. **Calculate how much energy it would take you to ride a bicycle to the store**, instead of walking or taking the car. Hint: A 180-pound person uses up about 100 calories per mile walking, while a 100-pound person uses about 53 calories per mile.
3. **Find a way to use leverage and friction to open jars that are really hard to open.**
4. **Explore better ways to move heavy household objects** like couches, pianos, and appliances.

Watch

To watch the recommended videos for this chapter, go to **generations.org/ GodMadeTheWorld** and scroll down until you find the video links for Chapter 6. Our editors have been careful to avoid films with references to evolution. However, we would still encourage parents or teachers to provide oversight for all internet usage. These videos may not give God the glory for His amazing creative work, so the student and parent/teacher should respond to these insights with prayer and praise.

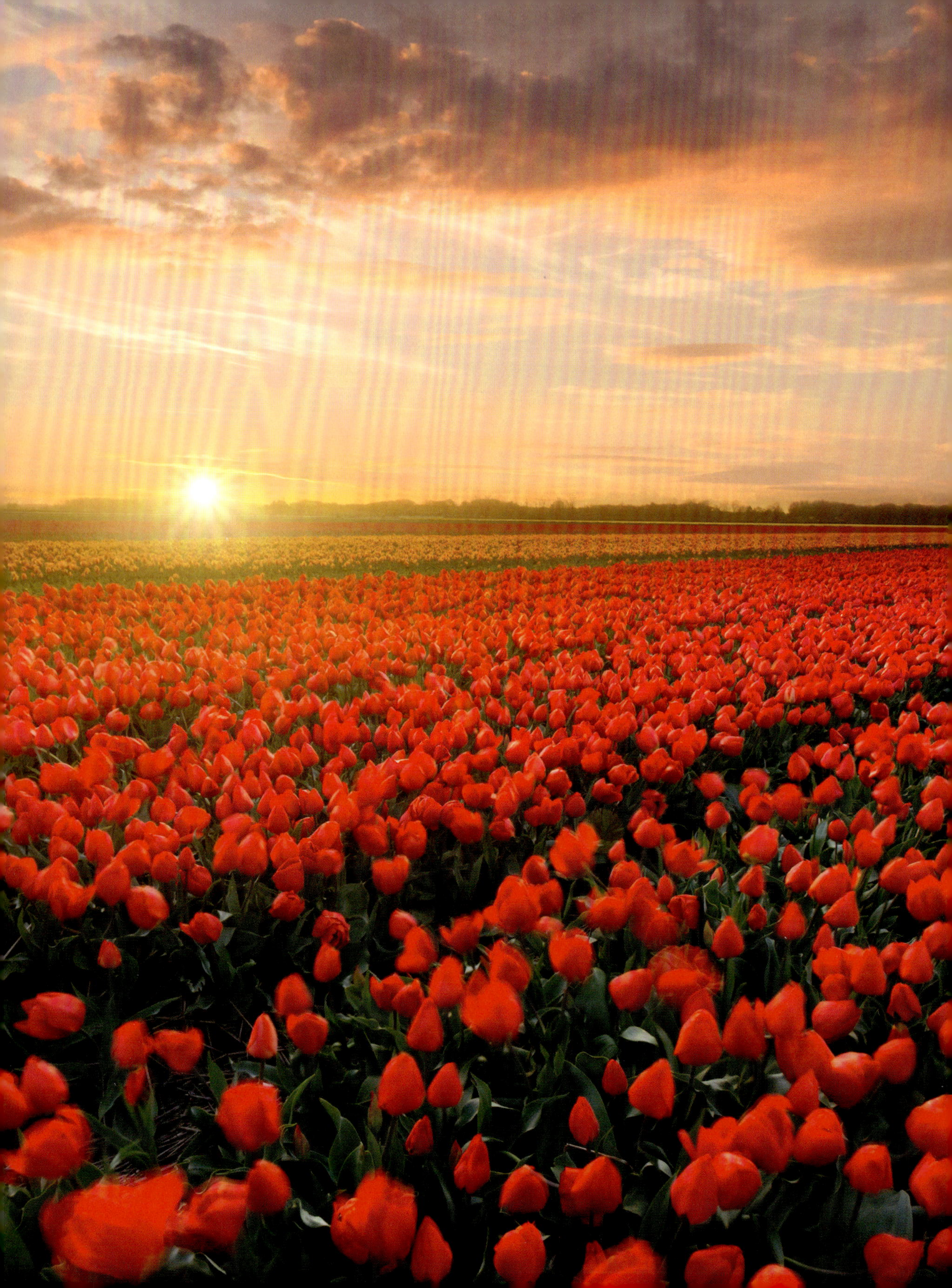

Chapter 7
GOD MADE WAVES—SOUND AND LIGHT

And God said, "Let there be light": and there was light. And God saw the light, that it was good: and God divided the light from the darkness. (Genesis 1:3-4)

Before God created light on the first day of creation, there was only darkness in the world. The most wonderful thing about light is its ability to overcome the darkness. Wherever the light shines, the darkness runs away.

Light is very mysterious. It is as baffling to scientists as the atom, gravity, and the electromagnetic force. All of these things reveal to us the very deep wisdom of God. He is much wiser than any of us, and we see this everywhere in His creation.

What is Light?

God is light, and in Him is no darkness at all. (1 John 1:5)

Scientists have a hard time defining light. We know that God is light, because that is what God's Word tells us. Those who study light believe it to be a stream of particles that flows like a wave. The particles are called **photons**. They have no mass. There's nothing to them, except energy. You cannot weigh a photon. You cannot cut a photon in two. Yet, a photon

GOD MADE THE WORLD

Street Lights

can "bump into" electrons which do have mass and weight to them.

Photons are not the same thing as matter, but they can be absorbed into matter. When light photons are absorbed into black asphalt roads, what happens to the roads? You don't want to walk across the road on a hot day in bare feet. The sunlight has energy in it and that is what heats up the road.

Light is another example of an electromagnetic wave. The scientist James Clerk Maxwell called light an **electromagnetic disturbance**. We know that changing magnetic fields will produce electric fields, and changing electric fields will produce magnetic fields. This is the most accurate description of what God is doing with the miracle of light. When the light energy is absorbed into the pavement, the photons disappear. When a light bulb turns on, or when a fire is lit, more photons are created as the light flows around the room. The most wonderful thing about photons is that they can bring energy all the way from the sun to the earth.

Here is a great mystery! What are these photons? They are not matter, but they bump into matter. This is another example of God's awesome creation. Once again, we are amazed at the creation of God. Thankfully, we can know a little bit about God and His creation. But we cannot completely comprehend it. Our minds are just too small. What more can we do than to praise Him and say,

Oh, the depth of the riches both of the wisdom and knowledge of God! How unsearchable are His judgments and His ways past finding out! (Romans 11:33)

The Uses of Light

In Your light we see light. (Psalm 36:9)

The most important use for light is

Mountain Biking with a Head Lamp

sight. Without light, we wouldn't be able to see the beauty of God's creation. In fact, sight is the most important of all human senses. We could not work without sight, and scientists could never discover God's creation and make good use of it. We would not be able to build houses, cars, airplanes, and cellphones without sight. We would never survive. We would run into things all the time, and we couldn't defend ourselves in dangerous situations. But most importantly, using the sight God has given us, we can see the glory, the power, and the wisdom of God's creation. When we look at the sun, the moon, the stars, sunsets, mountains, flowers, parrots, and man, we see God's glorious handiwork. And, we witness all of these things expressing His glory.

It is only in the light provided by God that we can see the truth about our world. We can understand a little of the world by the light of the sun. But mostly, we understand all things by the light of God's Word. His Word is our light.

GOD MADE THE WORLD

Oldest Surviving Photograph Taken by Joseph Nicéphore Niépce

> And so we have the prophetic word confirmed, which you do well to heed as a light that shines in a dark place, until the day dawns and the morning star rises in your hearts. . . (2 Peter 1:19)

Light also helps us to see everything in color, and here again we experience God's goodness and glory.

Cameras use light to record a picture. By this means, we can keep a visual picture as a record of something that happened in the past. Before photographs, artists would make sketches and paintings. But now, you can record thousands of pictures of your family and things going on in your home throughout the day. The first photograph was taken in 1826 by a French inventor named Joseph Nicéphore Niépce.

Compact disc players use light to play songs, and photocopiers and computer printers use light to make copies. Since light is energy, powerful lasers use a special kind of light to cut through steel. Fiberoptic cables use light as a means of carrying messages. These light signals bring computer data or phone calls (human voices) over long distances.

Although humans have invented many uses for light, the best and most useful light remains the sun. Even the moon is reflected sunlight, which helps us to see at night. There is nothing more glorious than light, because that is how God displays His glory. At Jesus's birth,

CHAPTER 7: GOD MADE WAVES—SOUND AND LIGHT

shepherds near Bethlehem witnessed God's singing glory, and they were struck with a reverential fear.

> Now there were in the same country shepherds living out in the fields, keeping watch over their flock by night. And behold, an angel of the Lord stood before them, and the glory of the Lord shone around them, and they were greatly afraid. (Luke 2:8-9)

Wave Energy

Some things travel in a straight line. But God has designed some things to travel in a wave. The water in the ocean sometimes travels in a wave. Drop a stone in water, and watch the ripples move in waves, one by one out from the center. Or you can tie a small rope to a door knob, and move the other end up and down rapidly. The rope will carry the energy in a wavelike motion.

Here are some terms you need to know about waves:

- **Amplitude**—A measure of the distance from the top (crest) of the wave to the bottom (trough).
- **Wavelength**—A measure of the distance between crests.
- **Frequency**—The number of waves that pass in one second.
- **Wave speed**—The speed of a wave in meters per second.

Wave speed is calculated using the following equation: $v = f\lambda$.

v is speed, f is frequency, and lambda is the wavelength.

The Electromagnetic Scale

All **electromagnetic energy** is made up of photons. Light is only one example of electromagnetic wave energy. The various kinds of electromagnetic energies are organized by wavelength in the following table.

If you were swimming at the beach,

Ripples in Water: Example of Wave Energy

Wave	Wavelength Width of wave in nanometers	Range of Frequency Number of waves per second
Radio Waves	>108	3 billion per sec
Microwave	105- 108	3 billion - 3 trillion per sec
Infrared	700 - 105	3 trillion - 430 trillion per sec
Visible Light	400-700	430 trillion - 750 trillion per sec
Ultraviolet Light	10-400	750 trillion - 30 quadrillion per sec
X-ray	1-10	30 - 300 quadrillion per sec
Gamma Waves	<1	Over 300 quadrillion per sec

you would probably feel the waves of water moving around you. The electromagnetic waves are constantly moving all around you. You can't feel them, and you can't see most of them. The only waves that can be seen are light waves. Electromagnetic waves with shorter wavelengths called **X-rays** and **Gamma Rays** are dangerous for people.

All of these waves travel at the same speed through empty space—186,000 miles per second (300,000 kps). Since there is no air in space, light travels very quickly.

When the light waves run into the earth's atmosphere, they slow down. Some waves will travel right through a human body, but not at the high speed at which they travel through space. Think about how much harder it would be to

CHAPTER 7: GOD MADE WAVES—SOUND AND LIGHT

travel through an ocean of honey than to fly through the air. Similarly, light finds it easier to travel through space than to travel through the air or water.

Also, the waves with the shorter wavelengths travel the fastest through air, water, and other solids.

Scientists can get rid of the air in a tube, by creating a vacuum. Without air, a vacuum is similar to space above the earth's atmosphere.

Speed of Light Traveling Through Various Things

Medium	Speed of Light
Vacuum	186,276 miles per second (299,790 km/second)
Air	186,220 miles per second (299,700 km/second)
Water	140,000 miles per second (225,000 km/second)
Glass	124,000 miles per second (200,000 km/second)

The Light Scale

And God said. . . "The rainbow shall be in the cloud, and I will look on it to remember the everlasting covenant between God and every living creature of all flesh that is on the earth." (Genesis 9:12,16)

Our Creator God brings light into the world with a wonderful color spectrum—each color with a different wavelength. For thousands of years, man could see this by the rainbow which God put in the sky. Then, in 1672, Isaac Newton discovered the light spectrum. He put a beam of light through a prism, and saw seven distinct color bands appear—just like the rainbow. God put seven bands into the rainbow, just as there are seven days in the week. It is the

Glass Cup

Light Spectrum

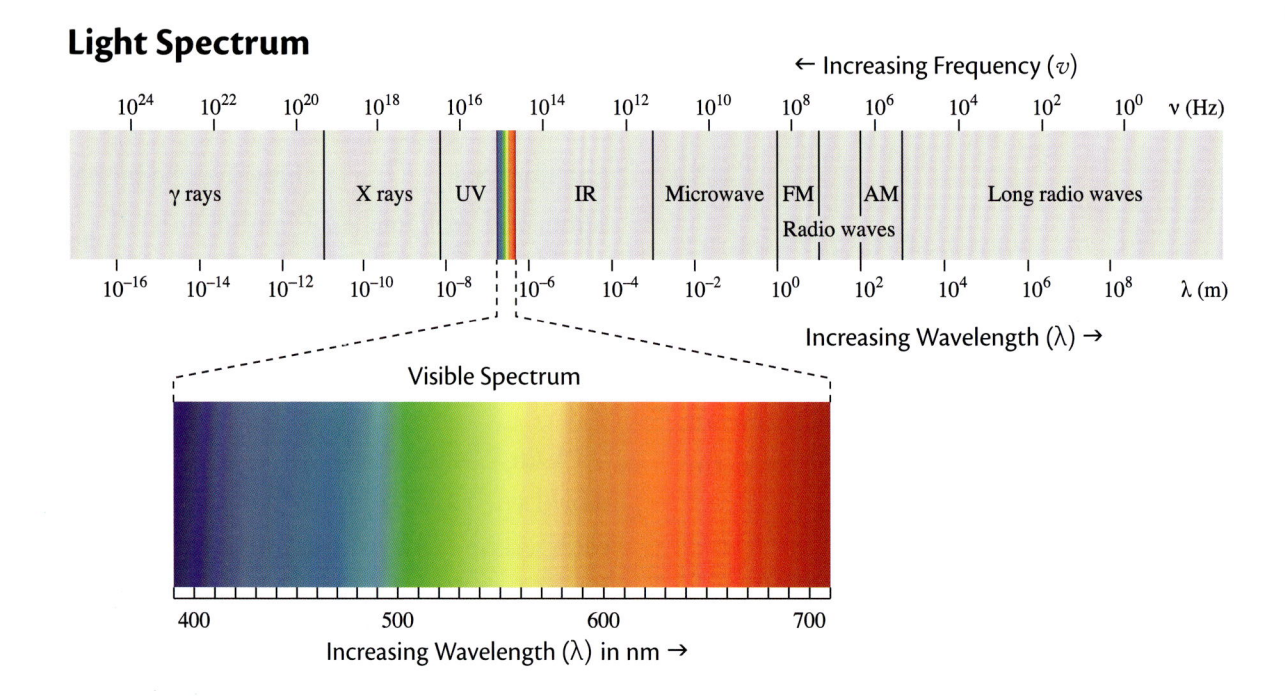

Visible Spectrum

Increasing Wavelength (λ) in nm →

number of perfection and completion. The seven colors are red, orange, yellow, green, blue, indigo, and violet. You can remember these colors by memorizing this fellow's name: Roy G. Biv. Each letter of his name stands for a color.

Light will travel through a medium (a substance like air, water, or glass). When it hits the substance, the light slows down, and it changes direction a little bit. This bending of the light beam is called **refraction**.

A material like plastic or glass is **transparent** if you can see through it. That's because the light is traveling straight through it without bending. If you can't see through it very well, the material is **translucent**. The light bends slightly as it enters, and only some of the light actually makes it through. Super thick curtains or window shades will block all the light. These are called **opaque** materials. They do not allow any light to come through. They either block the light or absorb the light into the material.

Light displays color by reflection and refraction. First, you need to know about refraction, which is the bending of a light ray. Each light wave carries with it a rainbow of colors, each with its own wavelength. When a light beam passes through a glass prism,

CHAPTER 7: GOD MADE WAVES—SOUND AND LIGHT

the wave bends. But each of the colors within the light beam bends at a slightly different angle and they separate from each other. The same thing happens with the rainbow. When the light beam shines through the raindrops, the light refracts and the colors separate into a rainbow.

Color Wavelengths

Light can also bounce off of surfaces. If you throw a rock into a pond, it will slow down, but it will also change direction in the water. But if you throw the rock at an angle, it may bounce off the water. Light operates in a similar way. Similar to the rock, light does different things when it collides with water, air, or solids. Some of the light may go all the way through the material, refracting or bending. Some of the light may absorb into the material and transfer into heat energy. Some of it bounces off. This is called **reflection.**

Color	Wavelength
Red	700 nanometers
Orange	600 nanometers
Yellow	580 nanometers
Green	525 nanometers
Blue	480 nanometers
Indigo	445 nanometers
Violet	435 nanometers

Rainbow over Manhattan, New York City

Colored Pencils

How Do You See Colors in God's Beautiful World?

Our Creator God has wonderfully designed light to react to certain colors. When light hits a black surface, almost none of it bounces off. Your eye does not see any part of the light beam bouncing off and that is why the surface appears black to your eye. All the wavelengths of light and all the colors in the wave are absorbed into the material, and the energy turns to heat. Dark surfaces like black t-shirts absorb more heat from the sun than white t-shirts. That's because most of the light beam is transferred into heat. None of the light is reflected.

If the surface is colored red, then only the red wavelength will reflect back into your eyes. The other wavelengths of orange, yellow, green, blue, indigo, and violet will be absorbed and disappear. If all the colors of the whole light beam are reflected, then you will see white. God made varying pigments, which are substances that absorb light at different wavelengths. Red pigment reflects red and absorbs the other wavelengths. Blue pigment reflects only the blue.

The primary colors for light are red, green, and blue. They are called **primary colors** because each one is pure and it is not mixed up into other colors. Mixing any two or three primary colors will produce

CHAPTER 7: GOD MADE WAVES—SOUND AND LIGHT

all the other colors. Incredibly, the human eye can distinguish one million shades of colors. But that's not all. Some people are super sensitive to color—they can tell the difference between 100 million shades. What great variety our Creator God has put into our world!

God created light, pigments, and the human eye to reveal a beautiful, colorful world. Without all three things available to us at the same time, we would only see the world in black and white! You can see the wisdom of God in this design. He made these things so that the human eye could see the beauty of His creation. He wants us to take pleasure in it, and then to thank Him for it. He wants us to be able to see the glory of His creation, and then give Him praise and worship in response.

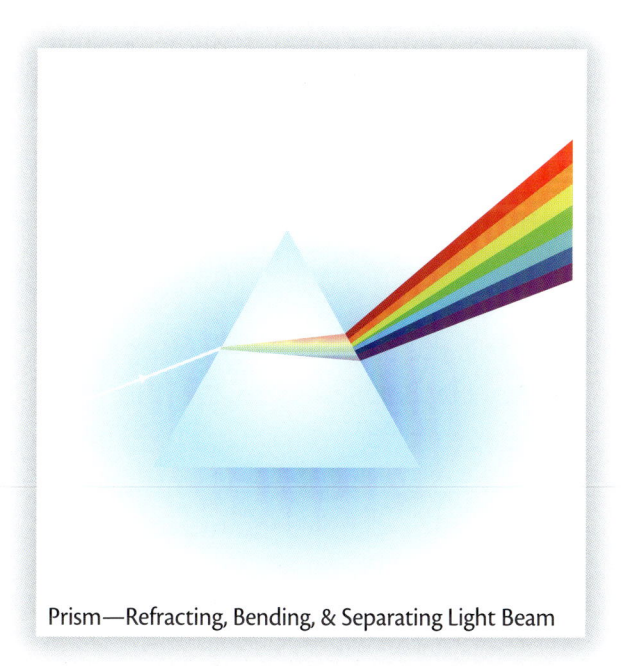

The heavens declare His righteousness,
And all the peoples see His glory. (Psalm 97:6)

Light Helps Us to See

Then Jesus said to them, "A little while longer the light is with you. Walk while you have the light, lest darkness overtake you; he who walks in darkness does not know where he is going." (John 12:35-36)

You cannot see anything without light. If you turn off the lights, and cover the windows with black paper, it would be pitch dark in the room. The things in the room do not disappear. All the furniture is still there. You still have eyes to see things. You have a brain which can interpret what your eyes can see. But, without light your eyes cannot see anything. God created your eyes, and He created light so you could see everything around you.

The light beams must reflect off of the

Prism—Refracting, Bending, & Separating Light Beam

Light Beams Traveling Through Mist

surfaces in order for you to see anything. When a light bulb is turned on in the room, the beam starts bouncing off of everything (almost all at once). Immediately, everything becomes visible to you.

If you were to shine a light beam from a small penlight into a mirror in a dark room, you wouldn't see much. The beam would reflect off the mirror onto a wall. You would only see a bright spot shining in the mirror and another bright spot on the wall. That's because light beams do not reflect off of walls very well. Now, if somebody were to spray a mist of water into the beam, you would see the light beam from the penlight traveling through the air. The water mist reflects a little bit of the light such that your eye could pick it up.

In John 8:12, Jesus said "I am the light of the world." Though light itself is a mystery, we do know the source of light. It is the Son of God, the Lord Jesus Christ. He enables us to see the truth about our world around us. He helps us by lighting up the physical world so we can see trees, people, mountains, rivers, and stars. But, He also helps us by giving us His Word.

CHAPTER 7: GOD MADE WAVES—SOUND AND LIGHT

Your Word is a lamp to my feet,
And a light to my path. (Psalm 119:105)

By His Spirit, Jesus shines light into our dark hearts, and helps us to understand this Word. We can come to understand the most important truths in the Bible. By His light, we can see that we are sinners, and that we need a Savior. We can see the truth and the beauty of His commandments. We can understand them. In His light, we can begin to walk in the right path, without stumbling. We can see the gloriousness of Christ. Then, we praise God our great Redeemer, our King, and our Creator! Without this light, we cannot see any of this. We could never know the truth. And we would never see the beauty and glory of Christ. We could never join in worship on the Lord's Day with joy and loud praise from the heart!

Mirrors

Mirrors are made to reflect light. Just like most things about light, we don't know exactly how reflection works. The photons are absorbed by the mirror, and

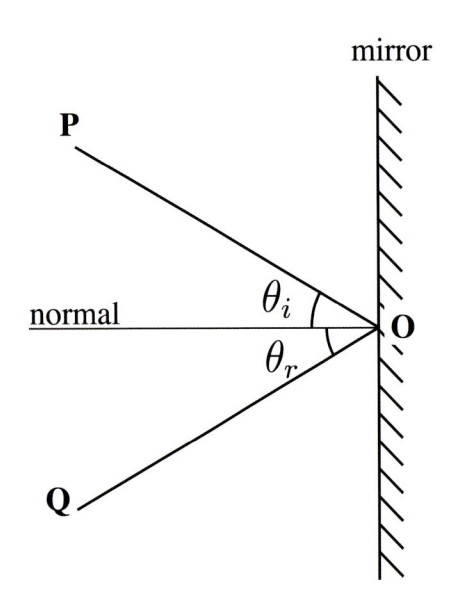

Diagram of Mirror Reflection

then other photons emerge back out of the glass at an equal speed.

When light rays hit a rough surface like concrete or paper, the light reflects in all directions. Mirrors, however, are made of a very smooth surface. Pretty much all the light coming in at one angle will reflect back towards us at the same angle. If a bunch of light beams come in at 90 degrees, they will all reflect at 90 degrees, straight back at us. If a light beam travels into the mirror at 45 degrees, it will reflect away at 135 degrees, as the figure above shows. The light beam bounces off the mirror at an equal angle, like a ball thrown at a wall at an angle.

But what is light doing in order that you can see a reflection of yourself in the mirror? Light rays are always moving quickly, everywhere, all around a lighted room. Light does not refract on solid objects like human bodies. Instead, it reflects, or bounces off your body. There are many rays bouncing off of your face, your nose, your cheeks, your hair, and your eyes whenever you are in a lighted room. Some of the thousands of rays hitting your face will travel into the mirror. And some of these rays will bounce back towards your eye. The light rays reflected at every point on your face will reveal different colors—hazel eyes, pink lips, dark cheeks, brown hair, and white teeth. Remember that each ray of light will reveal the wavelength(s) reflected off your face.

Your image will look like it sits behind the glass. For a flat mirror, the image will look like it is as far behind the mirror as you are standing in front of it.

Converting Light Energy into Electricity

One of the practical inventions that came by science was photo-electric cells. When photons from light bump into a bunch of atoms, they can knock

CHAPTER 7: GOD MADE WAVES—SOUND AND LIGHT

Albert Einstein (1879-1955)

How Solar Cells Work

It is the glory of God to conceal a matter, But the glory of kings is to search out a matter. (Proverbs 25:2)

Solar cells are made out of silicon. This material is neither an insulator nor a conductor. It is a semi-conductor.

A godly, Christian scientist named Michael Faraday was the first to discover the semi-conductor in 1833. All computers and almost every electronic device today have been developed out of Michael Faraday's discoveries.

He found that some materials increase their conductivity with an increase in temperature. These materials, called **semi-conductors**, also increase their conductivity when certain impurities are mixed into it. This is called **doping**.

When you dope silicon with some stuff like arsenic or phosphorus, the silicon will gain extra electrons. This is called the **n-type silicon**. Mixing boron or aluminum into the silicon creates a material with fewer electrons. This is the p-type silicon, and it will be more likely to receive electrons.

For a solar cell that will create elec-

the electrons out of place. This creates electricity from solar energy. A humble and brilliant scientist named Albert Einstein discovered this process, and received the Nobel Prize for Science for his discovery in 1921. Actually, sunlight carries a lot of energy with it, an average of 164 Watts for every square meter. If an energy company would put solar panels on 1% of the Sahara Desert, that would provide enough energy for the whole world.

Solar Panels

tricity, you must make a sandwich of an n-type silicon piece on top of a p-type silicon piece. So far, there is no electricity flowing. You have not produced any energy by putting this material together.

However, if you shine a light beam of photons on the cell, the light energy is changed to electrical. The photons start moving electrons in the n-type piece creating an electrical charge between the top and bottom layers. When electric wires are connected to both sides, a circuit is created and electrical energy begins to flow.

God Made Sound Waves

The hearing ear and the seeing eye,
The LORD has made them both. (Proverbs 20:12)

God provided man with sights, light, and eyes to see, but that's not all! Praise be to God for ears, made to hear the many sounds He has created. He also gave us vocal cords and breath so we could make sounds and speak to one another.

It takes energy to create sound. As with light, sound travels in waves. The sound wave itself is considered mechanical

CHAPTER 7: GOD MADE WAVES—SOUND AND LIGHT

Sound Wave

Michael Faraday

The great scientist, Michael Faraday, who was the most helpful in developing the field of electronics, wrote, "The book of nature which we have to read is written by the finger of God."[1]

He wrote to a friend in 1861, urging him to trust in Christ: "Since peace is alone in the gift of God; and since it is He who gives it, why should we be afraid? His unspeakable gift in His beloved Son is the ground of no doubtful hope."[2]

Faraday was on the brink of death in 1867, and his wife asked him, "What will be your occupation in heaven?" His response was: "I shall be with Christ, and that is enough."[3]

Michael Faraday (1791-1867)

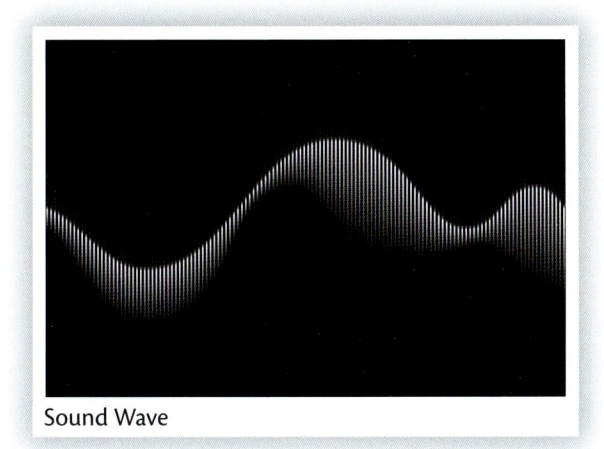

energy—an energy that vibrates an object. The ticking sound of a clock comes by moving parts. Cymbals clashing together cause vibration. This produces a sound wave that travels through the air. These waves can travel through gases (like air), water, and solids. If somebody is making a racket in the room next to you, the sound may travel through the walls. Sound waves move out from the source of the sound in every direction.

The ear is a marvelous invention of our Creator. Sound waves hit the eardrum and cause it to vibrate. Nerve endings in your ear take these messages to your brain, and your brain figures out the message. Was it a voice? Was it a loud crash? Was it a dog barking? We learn what most sounds mean by experience, from the time we are

Loudspeaker—Electric Energy Moves Voice Coil

very young. When we see the dog barking and hear the sharp sound, we begin to associate the sound with the animal. So the next time we hear the sound, we interpret it to be a dog barking even if we cannot actually see the dog with our eyes.

God's creation of hearing is very complex and capable of so much! Hundreds of sound waves can surround you at the same time, and your ear can sort them out. The dog barks, the door slams, the music plays, and the refrigerator compressor hums along, all at the same time. You can listen to a symphony and enjoy one hundred instruments playing at the same time.

How does a sound wave work? The sound wave travels more like a spring or a "slinky." When the vibrating sound wave forces particles in the air or medium together, this is called **compression**. When the vibrating waves spread them apart, this is called **rarefaction**.

Sound travels a lot slower than light—about 335 meters per second in air. Amazingly, sound travels 4.3 times faster in water than it does in air. That is because sound travels faster through dense material, and water is more dense than air. Unlike light, sound cannot travel through outer space, because it needs to push through a medium (like air, water, or other things).

The Speed of Sound Through Various Kinds of Materials	
Air	335 m/sec
Water	1,500 m/sec
Oak Wood	3,900 m/sec
Iron	5,955 m/sec

There are several terms you need to know relating to sound, some of which are similar to light waves.

Frequency is the number of full waves that are traveling through a point (or hitting your ears) in one second. This is measured in Hertz. Twenty hertz is twenty full cycles passing in one second.

Wavelength is the length of one wave. How quickly does the source of sound vibrate? A very long wavelength is a very slow vibration. A short wavelength vibrates faster.

Pitch is related to frequency. A woman's scream has a high frequency. A bass drum vibrates at a low frequency. The higher the frequency the higher the pitch. When the motor of a mixer runs faster, the frequency is higher and so is the pitch. At the very highest setting, the motor sounds like it's screaming at you. If you were to turn it down, the sound takes a lower pitch.)

In God's great wisdom, He designed His creatures with the ability to hear (pick up) different ranges of sound frequencies. See the chart on the following page.

Most animals can hear sounds at frequency levels much higher than what humans can hear. Cows, elephants, goldfish, and cats can hear sounds at lower pitches than humans can.

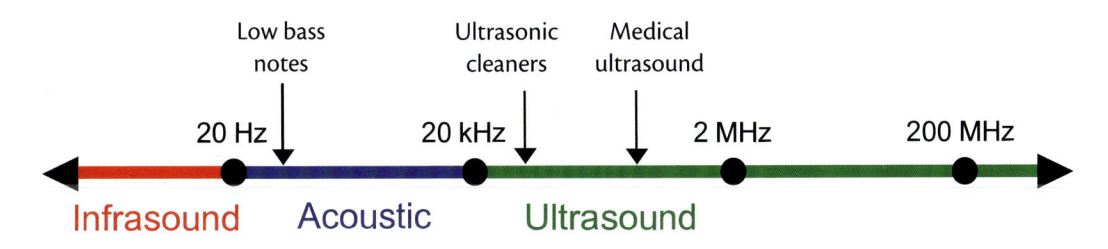

Low bass notes — 20 Hz
Ultrasonic cleaners — 20 kHz
Medical ultrasound — 2 MHz
200 MHz

Infrasound | Acoustic | Ultrasound

Frequency of Sound

Species	Approximate Range (Hz)	Species	Approximate Range (Hz)
Human	64-23,000	opossum	500-64,000
Dog	67-45,000	bat	2,000-110,000
Cat	45-64,000	elephant	16-12,000
Cow	23-35,000	porpoise	75-150,000
Horse	55-33,500	goldfish	20-3,000
Sheep	100-30,000	catfish	50-4,000
Rabbit	360-42,000	bullfrog	100-3,000
Rat	200-76,000	tree frog	50-4,000
Mouse	1,000-91,000	canary	250-8,000
Guinea pig	54-50,000	parakeet	200-8,500
Raccoon	100-40,000	owl	200-12,000
Ferret	16-44,000	chicken	125-2,000

CHAPTER 7: GOD MADE WAVES—SOUND AND LIGHT

Volume

Another measure of sounds is volume, or loudness. Volume is measured in **decibels**, with 0 decibels being the lowest point at which the human ear can detect sound. Normal conversations are about 60 decibels. The sound of a jet engine at takeoff will exceed 160 decibels.

Microphones—Converting Sound Energy to Electric Energy

One of the most important purposes for science is to find out new ways to change one form of energy into another. The microphone is a good example of this. When you speak into a microphone, the sound waves from your voice vibrate on a diaphragm (sort of like a small trampoline) in the mic. The diaphragm is attached to a coil of wire wrapped around a magnet. The coil of wire vibrates up and down, creating an electric current. The electric signal is patterned after the sound waves coming from your speaking voice. This electric signal may be recorded or amplified through speakers.

Speakers—Converting Electric Energy into Sound Energy

To broadcast or amplify sound through a public address or loud speaker system, the electric energy coming through the microphone needs to be changed back into sound energy. If you look into the back of a speaker, you will find an electric coil wrapped around a flexible electromagnet. Every time the electric signal comes through the wire, the electromagnet is magnetized. A permanent magnet sits behind it. The electromagnet begins to vibrate back and forth towards the permanent magnet as the electric signal passes through the wire. The electromagnet is attached to a drum-like paper diaphragm, which also moves in and out. This movement pumps sound waves into the room, which makes the sound much louder, especially if the electromagnet is moving vigorously (powered by lots of electrical energy).

Microphone

GOD MADE THE WORLD

In short, the energy transfer works in this way:

Microphone:
- Sound energy changed to mechanical energy of the diaphragm
- Then mechanical energy changed to electrical energy

Speaker:
- Electrical energy changed to mechanical energy
- Then mechanical energy changed to sound energy (sound waves)

How Loud is Too Loud?

Many young people like to listen to loud music at concerts and in church. A recent test of eight churches found the volume to be between 93 and 98 db. Some rock concerts run the music up to 130 db. The human ear could be damaged if it takes in 115 decibels for more than 30 seconds.

Sound waves at certain frequencies can cause huge vibrations in solids and liquids. This is called the **resonant frequency**. This is what causes a wineglass

Dolphin Use Echo-Location to Gage Distances

to break under certain sounds. When the sound frequency is set at a level at which the glass would naturally vibrate, and assuming the sound is set at a high enough volume, the glass vibrates faster and faster until it breaks.

Some birds use echo location to find their way in the dark. For example, the oilbird flies around caves making a clicking noise at 7000 Hz. The sound bounces off of a solid surface, and creates an echo. The bird then calculates the amount of time it takes for the echo to make it back. From that, he can figure out the distance of the objects around him. This high-tech process, called **echo-location**, is another phenomenal design of our Creator for birds, bats, whales, and dolphins.

Doctors and nurses use ultrasound to look at a baby inside his mother's womb. The ultrasound machine produces a bunch of high frequency sound waves, and then calculates the time it takes for the echo, or sound wave, to bounce back to the probe. As millions of sound waves travel through various parts of the woman's and baby's bodies, they speed up or slow down depending on the type of fluid or tissue that they encounter. Then the machine analyzes the speed and time of the sound waves returning. Based on this information, an image is created of the little baby in the womb.

Acoustics

Acoustics is the study of sound. Similar to light, sound will bounce off of some surfaces and be absorbed into others. Sound will absorb into uneven surfaces better than smooth surfaces. Sound waves can't bounce directly off soft, fluffy, porous, or rough surfaces.

Sometimes you might notice an echo in an empty room. This happens when there is too much sound bouncing from wall to wall. You can reduce the echo quite a bit by installing carpets on the floor, and curtains on the windows. Even wallpaper can absorb some of the sound.

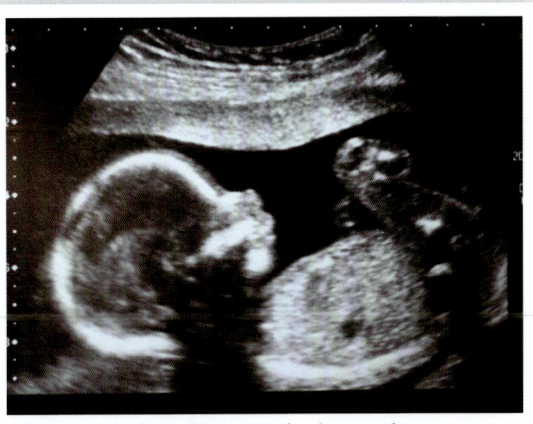

Ultrasound of a Baby in Mother's Womb

GOD MADE THE WORLD

Also, adding furniture and putting throw blankets over the furniture can help.

Communicating with Sound

Shout joyfully to the LORD, all the earth;
Break forth in song, rejoice, and sing praises.
Sing to the LORD with the harp,
With the harp and the sound of a psalm,
With trumpets and the sound of a horn;
Shout joyfully before the LORD, the King.
Let the sea roar, and all its fullness,
The world and those who dwell in it;
Let the rivers clap their hands;
Let the hills be joyful together before the LORD. . . (Psalm 98:4-8)

God created sound, and He has commanded us to make sound. He wants to hear our praise. He wants us shouting joyfully, singing, and clapping in His worship. He wants to hear our expressions of joy over His salvation, for He has saved us from the most horrible condition of sin and death. Our LORD Jesus got a great victory for us by His cross and resurrection! We use sounds to express joy and celebration.

Immediately after creating Adam, the Lord God spoke to him. He came into covenant relationship with the man by way of the spoken word. He told Adam not to eat from the tree of the knowledge of good and evil, or he would die. Men and women relate to God and to each other by the spoken word. We build relationships by speaking to each other. Our words are really powerful. They can be used to build up one another, or they can be used to cut down.

Indeed, we put bits in horses' mouths that they may obey us, and we turn their whole body. Look also at ships: although they are so large and are driven by fierce winds, they are turned by a very small rudder wherever the pilot desires. Even so the tongue is a little member and boasts great things.
See how great a forest a little fire kindles!

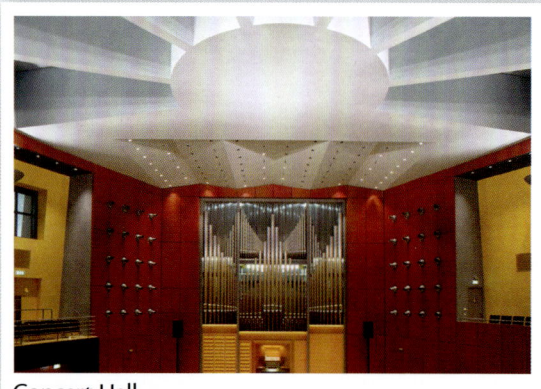

Concert Hall

CHAPTER 7: GOD MADE WAVES—SOUND AND LIGHT

God Gave Us Beautiful Music

And the tongue is a fire, a world of iniquity. The tongue is so set among our members that it defiles the whole body, and sets on fire the course of nature; and it is set on fire by hell. (James 3:3-6)

Music

Praise the LORD!
Praise God in His sanctuary;
Praise Him in His mighty firmament!
Praise Him for His mighty acts;
Praise Him according to His excellent greatness!
Praise Him with the sound of the trumpet;
Praise Him with the lute and harp!
Praise Him with the timbrel and dance;
Praise Him with stringed instruments and flutes!
Praise Him with loud cymbals;
Praise Him with clashing cymbals!
Let everything that has breath praise the LORD.
Praise the LORD! (Psalm 150)

God made music for His worship. He has commanded us to sing, and to play beautiful music in His worship. Music expresses our deepest thoughts, our joys, our hopes, and desires. Music is wonderfully pleasant, and it can be a comfort to those who are going through hard times. Music can affect our moods. It can prepare us for battle, encourage and strengthen us, and move us to love God and others. It can bring back memories. It can make us joyful or it can make us sad.

Music is another gift from our God, made possible by the workings of our hearts, our minds, our ears, our voices, our mouths, and our hands. Let us praise Him then! Let us thank Him every day. For all of this was created by God for our benefit and for His glory. ∎

Pray

- Let us praise the Lord for the mystery of light. Praise God for how the light chases away the darkness. Praise Him for the gloriousness of light, as a peek into His glory.
- Praise Him for creating light, pigments, and the human eye—all three of these. Thank Him for the ability to see beautiful colors, sunsets, fields of gorgeous flowers, and the beautiful seas.
- Thank Him for the light of His Word. Thank Him for Jesus, who is the Light of the world.
- Thank Him for beautiful sounds, and the ability to hear. Let us thank Him for the ability to speak and to carry on relationships with family and friends. Thank Him for the birds that sing, and the beautiful music we enjoy each day.

Sing

The Light of the World is Jesus

The whole world was lost in the darkness of sin,
The Light of the world is Jesus!
Like sunshine at noonday,
His glory shone in;
The Light of the world is Jesus!

Refrain:
Come to the light, 'tis shining for thee;
Sweetly the light has dawned upon me;
Once I was blind, but now I can see:
The Light of the world is Jesus!

No darkness have we who in Jesus abide;
The Light of the world is Jesus!
We walk in the light when we follow our Guide!
The Light of the world is Jesus!

Ye dwellers in darkness with sin-blinded eyes,
The Light of the world is Jesus!
Go, wash at His bidding, and light will arise;
The Light of the world is Jesus!

No need of the sunlight in Heaven we're told;
The Light of the world is Jesus!
The Lamb is the Light in the city of gold,
The Light of the world is Jesus!

CHAPTER 7: GOD MADE WAVES—SOUND AND LIGHT

Do

1. **How many shades of color can you see?** Test your eye's ability to distinguish colors using an online test like https://www.igame.com/eye-test/.
2. **How have we discovered the usefulness of God's waves?** How might you find uses for these waves in your own life? Make a list of the uses of the following waves:
    a. Radio wave
    b. Microwave
    c. Infrared
    d. Ultraviolet
    e. X-ray
    f. Gamma ray
3. **You can always have power for your communication devices even when the electricity goes out.** Build a simple solar-powered USB/smart phone recharger using a kit purchased online. Kits are available from the Kitables company.
4. **Be sensitive to the noise pollution you are sharing with your neighbors.** How well does the insulation in your home keep the sound from escaping through the walls? You don't want your neighbors bothered by sounds. With the permission of your parents, turn on a radio or music source in the house. Test the decibels using a db sound meter. At what point (at how many decibels) are you able to pick up the noise just outside your front door? How loud must the sound be in order to pick it up at your neighbor's property or house?

Watch

To watch the recommended videos for this chapter, go to **generations.org/ GodMadeTheWorld** and scroll down until you find the video links for Chapter 7. Our editors have been careful to avoid films with references to evolution. However, we would still encourage parents or teachers to provide oversight for all internet usage. These videos may not give God the glory for His amazing creative work, so the student and parent/teacher should respond to these insights with prayer and praise.

Lake Tahoe, California, USA

Chapter 8

GOD MADE DIRT AND WATER

> Then God said, "Let the waters under the heavens be gathered together into one place, and let the dry land appear"; and it was so. And God called the dry land Earth, and the gathering together of the waters He called Seas. And God saw that it was good. (Genesis 1:9-10)
>
> And God said, "See, I have given you every herb that yields seed which is on the face of all the earth, and every tree whose fruit yields seed; to you it shall be for food. Also, to every beast of the earth, to every bird of the air, and to everything that creeps on the earth, in which there is life, I have given every green herb for food"; and it was so. (Genesis 1:29-30)

When God created the earth, He provided air for us to breathe. But we need more than air to live. We need food and water. Survival on other planets is not possible because these rocky, barren places lack the necessary things for life. In this chapter, we will look more closely at some of the stuff all around us—dirt and water. These are very important for human life, animal life, and plant life on earth.

Most of the world is made up of water. Oceans and lakes cover 70% of the world's surface. Most of it is salt water, with fresh water making up only 2.5% of the world's water. If all the water in the earth could

Greenland

be squeezed into 100 one-liter bottles, 97 bottles would be filled with ocean water. Two bottles would be frozen water found mostly in the Arctic and the Antarctic. The last liter would be all the fresh water lakes, rivers, and ground water in the world.

Water is very necessary for life on earth. In fact, most people would die within three days without water. The longest anybody has ever gone without water was eighteen days. Austrian police put a fellow in a prison cell and forgot all about him. He was barely alive when they found him eighteen days later.

Jesus said, "He who believes in Me, as the Scripture has said, out of his heart will flow rivers of living water" (John 7:38). And He told the woman at the well that He could provide her with water that would last forever. She liked the idea of never being thirsty again, but He was talking about a life-giving water for the soul. When He shares this with us, that water becomes a spring in our own hearts that provides for eternal life. He said, "Whoever drinks of the water that I shall give him will never thirst. But the

CHAPTER 8: GOD MADE DIRT AND WATER

water that I shall give him will become in him a fountain of water springing up into everlasting life" (John 4:14).

The Miracle of Water

As with most fluids, gases, and solids, when water heats up, it expands. The heat causes the electrons in the atoms to get excited, and they start moving really fast. As they move faster, they need more space and the material actually expands. When things cool down, the materials will contract or shrivel up. Water will contract as it cools, until it hits 39 °F (4 °C). Then, something very strange happens, which scientists cannot explain. As water begins to freeze into ice, it starts expanding and it becomes lighter than water. That's why ice floats on top of the water. What would happen if ice was heavier than water? Of course it would sink to the bottom. Then what would happen to the fish and the aquatic plants if the frozen water stayed at the bottom of the lakes and ponds? Of course, the plants would die and the fish could not feed on the plants. This ice floating on top provides insulation, to keep the water underneath warm so that the lakes and oceans do not freeze all the way through. This keeps the fish alive through the winter months.

"The LORD possessed me [wisdom] at the beginning of His way,
Before His works of old.
I have been established from everlasting,
From the beginning, before there was ever an earth. . .
When He established the clouds above,
When He strengthened the fountains of the deep,
When He assigned to the sea its limit,
So that the waters would not transgress His command,
When He marked out the foundations of the earth,
Then I was beside Him as a master craftsman;
And I was daily His delight,
Rejoicing always before Him. . . "
(Proverbs 8:22-23,28-30)

Boats and Buoyancy

One of the most useful purposes for water has been to provide a way of moving people and stuff from one place to another. Long ago, when Noah's sons and grandsons began to build the first towns, they put them on rivers. Before building roads, which takes a lot of time and hard work, people would carry their goods to city markets on the rivers. And still to this day, 90% of the world's goods are shipped

231

Buoyant Force

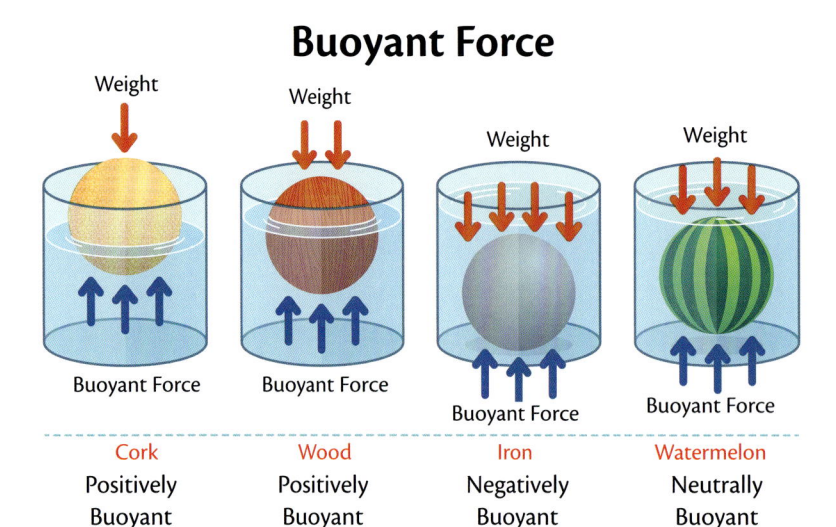

Cork	Wood	Iron	Watermelon
Positively Buoyant	Positively Buoyant	Negatively Buoyant	Neutrally Buoyant

on boats and barges.

Now some things will float on water and some things won't. For example, rocks and coins will sink to the bottom of a pond. But things made out of wood and plastic will usually float. Why do some things float, and other things just sink?

Some materials and objects are more **buoyant** than others. Buoyancy is the ability to float. When you put something in the water, the weight and the shape of the object will push down on the water. Meanwhile, all the water molecules under the object are pushing against it, as if they were trying to keep it afloat. When you set a boat in a lake, it will sink a little way into the water. This is called **displacement**. This is a measurement of the amount of water pushed out of the way when the boat is placed into the water. An object sinks when the force applied by the object against the water is greater than the water's force pushing against it from underneath.

Now here are the things which help objects to float on water:

1. The material's density is the most important factor. For example, as we already mentioned, ice is less dense than water, and it floats. Sailors use iron anchors to keep their boats from drifting. Since iron is very dense, it will sink to the bottom of the sea where it can grab onto a rock.

2. The shape of the object is also key. If you roll some modeling clay into a ball and put it in the water, it will sink. But if you flatten it out into the shape of a boat, it will float.

Kayaking

If you took a gigantic 100-foot (30 m) diameter ball of metal and dropped it in the sea, it would just sink to the bottom. Now if you took that metal ball and flattened it all out into a big boat, four hundred feet long with one-inch thick metal, do you think it would sink? Many large shipping vessels are made out of iron and steel.

A lot of people can float on water when they are lying flat. That is because the density of the human body can be slightly less than the water. However, you cannot stand on water, and you cannot walk on water for two reasons.

1. There is too much displacement caused by all your weight placed on your feet. There is too little contact area between your body and the surface of the water. So, there is not enough buoyancy for the human body when it is in the standing position.

2. There isn't enough **surface tension** between your feet and the water, which causes your body to break through the surface of the water very easily. Water has a stickiness to it, allowing some creatures like water striders to skate across the water without sinking. This is due to the two hydrogen atoms which have two electrons that attract a positively-charged surface. You can

Container Ship

usually float a paperclip on water, especially if you rub it on your hand first, and it picks up a little grease.

The Lord of Creation Walked on Water

Only the Master of all creation could make it possible to walk on water. Only God can change these physical properties to make that happen. So, in the Gospel of Matthew, we read that Jesus walked about three or four miles across the Sea of Galilee. Then, He invited Peter to walk with Him. This outstanding miracle was one more indication that this was the Creator who made the water and the human body too. Since He made the water, He could change the properties of it so He could walk over it. Jesus even commanded the winds and the waves, and they obeyed Him. That is because He is the Lord of nature.

Now in the fourth watch of the night Jesus went to them, walking on the sea. And when the disciples saw Him walking on the sea, they were troubled, saying, "It is a ghost!" And they cried out for fear.

CHAPTER 8: GOD MADE DIRT AND WATER

But immediately Jesus spoke to them, saying, "Be of good cheer! It is I; do not be afraid." And Peter answered Him and said, "Lord, if it is You, command me to come to You on the water."
So He said, "Come." And when Peter had come down out of the boat, he walked on the water to go to Jesus. But when he saw that the wind was boisterous, he was afraid; and beginning to sink he cried out, saying, "Lord, save me!"
And immediately Jesus stretched out His hand and caught him, and said to him, "O you of little faith, why did you doubt?" And when they got into the boat, the wind ceased. (Matthew 14:25-32)

Water Pressure

Fluids and gases will flow under pressure. When water flows into your house, the pressure is the force pushing the water through the pipe. Most homes have between 45 and 80 pounds per square inch (psi) of pressure (276-552 kPa) pushing the water through the pipes.

Sea of Galilee

To get an idea of the force of 80 psi through a one-inch pipe, think about an 80-pound boy jumping on a pogo stick. That is the force of water coming through the pipe. But, 80 psi pouring through a three inch pipe would be like a 240-pound man standing on a pogo stick.

Most fire hoses use 150-300 psi of pressure. It takes about 300 psi to put out a fire on the 30th floor of a building. That kind of pressure moving through a fire hose could push a grown man down. Before you turn on the faucet, water pressure is pushing against the valve. When you turn it on, you will feel the water gush out on to your hand. You can feel the force or the pressure of the water on your hand.

Usually, pumps are used to push the water out of the earth into homes. Sometimes, cities will pump water up into a water tower instead. These towers are designed to use gravity to create a constant pressure in the water pipes for the water flowing into homes. (For every foot above the ground, the tower provides 1/2 psi. So, a 160-foot water tower provides about 80 psi of pressure.)

If you had a small water tower on the top of a three-story house, the water on the third floor would only flow at about five psi. The water on the second floor would flow at about ten psi. And the water on the bottom floor would flow at

Firefighter with Fire Hose

about 15 psi. Showers would be a lot nicer on the bottom floor of the house, because the water would be flowing out harder and faster. That is because there is so much more pressure provided by gravity from the water sitting at the top of the three-story house.

Underwater Pressure

Gravity also creates water pressure which increases under the deep seas. For every 33 feet (10 m) you sink under the water, the water pressure increases by 14.7 pounds per square inch. That's double the pressure your lungs get at sea level, when you are standing out of the water. If you were to hold your breath and dive to 33 feet, your lungs would shrink to half the size. Whales can descend as far as 7,000 feet deep, because God made them with flexible ribs and lungs that can contract and expand easily.

Scuba divers have a problem, though. The air they breathe from a tank into their lungs is equal to the pressure under the water. If the scuba diver dives 99 feet under the water, the pressure of air coming into his lungs is four times what it would be on earth. As a diver descends, the pressure increase causes the air in his body to compress. Air spaces in their ears and lungs and other places become like vacuums. What happens when soda factories put high pressure gas into a liquid? The gas stays mixed into the liquid until you open the bottle. Then, all that pressurized gas tries to escape and bubbles rush to the top. Something similar happens with divers. They breathe pressurized gas when they are deep underwater. If they come up to the surface too quickly, all that gas turns into bubbles, and tries to escape from their bodies. When those bubbles form in muscle tissue and blood it can be really painful. This condition is called the **bends** and it can be deadly. Scuba divers can avoid the bends by rising up to the surface slowly. They need to take a three-minute break every twenty feet.

Water for Life

And he showed me a pure river of water of life, clear as crystal, proceeding from the throne of God and of the Lamb. In the middle of its street, and on either side of the river, was the tree of life, which bore twelve fruits, each tree yielding its fruit every month. The leaves of the tree were for the healing of the nations. (Revelation 22:1-2)

Scuba Diving

God created water to sustain our lives, as well as all animal life and plant life on earth. Revelation 22 tells us that in heaven there will be another life-sustaining river of living water for us.

So much of the provision of food in our world comes automatically. Sunlight is the energy source most essential for plant life. God's rain also provides the water needed for plant growth. Only about 20% of farms and ranches use irrigation from water pumped out of wells and rivers. In other words, God provides most of the energy, the water, and the chemicals necessary for plants to grow. Although farmers plant the seeds and harvest the fruit, God has wonderfully provided an automatic supply of everything needed for plants to grow.

There are several reasons why plants need water. First of all, water carries nutrients from the roots up into the leaves. It does this through capillary action, a very mysterious process whereby fluids can flow upwards against the force of gravity. Water can actually move up a porous material by adhesion, cohesion, and surface tension. Another way to put

CHAPTER 8: GOD MADE DIRT AND WATER

this is that water can move when the molecules are more attracted to a surface than they are to themselves. When you step into a deep puddle, have you ever noticed how the water creeps further up your pants? Some giant redwood trees need 150 gallons a day to survive. Some of that water is absorbed into the leaves from fog. But, much of the water is drawn up into the tree by **capillary action**.

Also, water is necessary for **photosynthesis**, a process by which God grows little seedlings into big plants and trees. The water that has crawled up into the branches and leaves combines with carbon dioxide in the air. The energy from the sunlight hits the mixture causing an amazing chemical reaction. This manufactures plant food or glucose (sugar), as well as oxygen. That's what makes plants green up, and helps the plants to grow.

For as the earth brings forth its bud,
As the garden causes the things that are sown in it to spring forth,
So the Lord GOD will cause righteousness and praise to spring forth before all the nations. (Isaiah 61:11)

Watering the Garden

GOD MADE THE WORLD

Photosynthesis

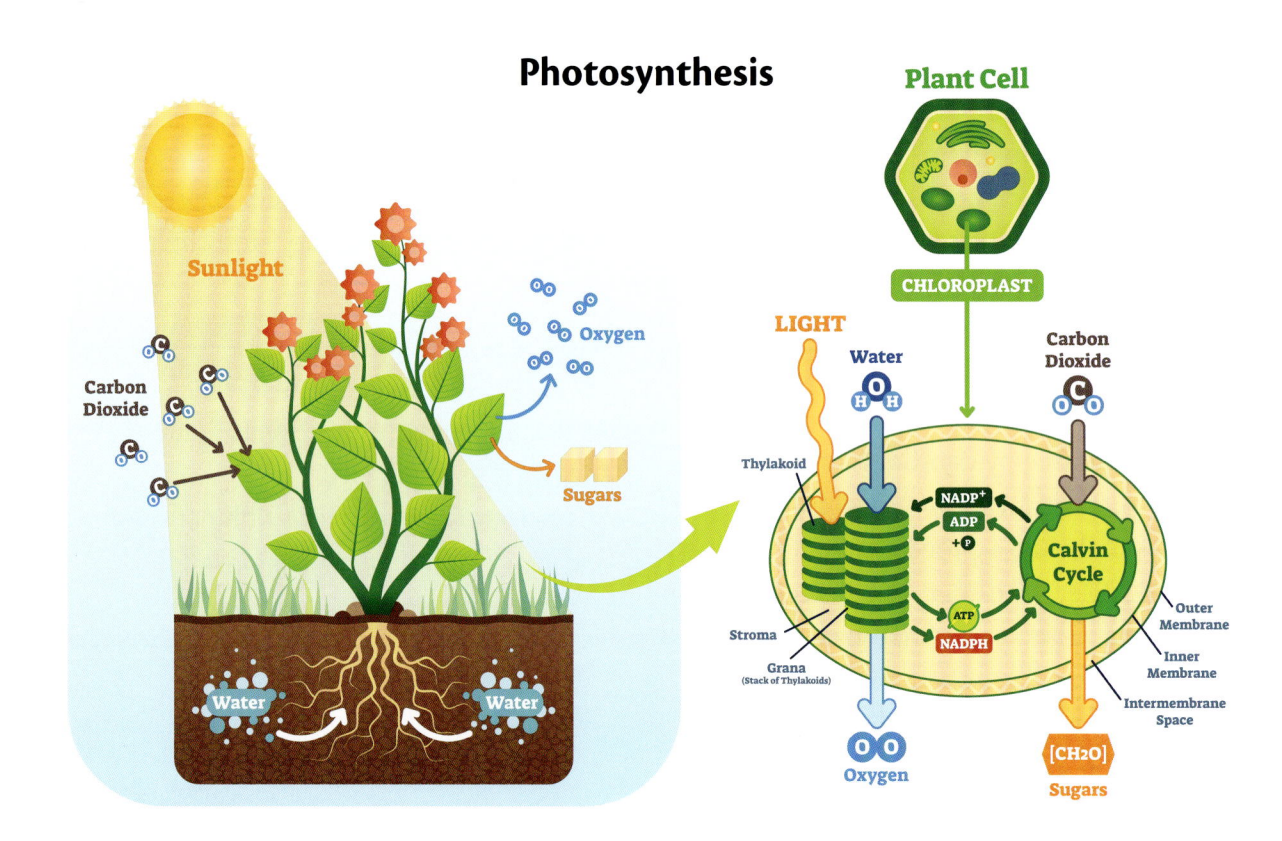

The Blessing of Snow

Then the LORD answered Job out of the whirlwind, and said:
"Who is this who darkens counsel
By words without knowledge?
Now prepare yourself like a man;
I will question you, and you shall answer Me...
Have you entered the treasury of snow,
Or have you seen the treasury of hail...
Has the rain a father?
Or who has begotten the drops of dew?
From whose womb comes the ice?
And the frost of heaven, who gives it birth?
The waters harden like stone,
And the surface of the deep is frozen."
(Job 38:1-3, 22, 28-30)

Snow is a wonderful creation of God. Normally, to get a solid out of gas, it needs to cool down and turn into liquid first. Then, it must be cooled down even

CHAPTER 8: GOD MADE DIRT AND WATER

more for it to turn into a solid. But God makes snowflakes (solids) such that they form directly out of gas (water vapor) in the clouds. Scientists have studied snowflakes for over one hundred years, and they still can't figure out how so many different shapes are formed. There are no two snowflakes exactly the same.

Wilson "Snowflake" Bentley (1865-1931) was the man who first studied snowflakes while he was being homeschooled by his mom. After reading about the "treasury of the snow" in Job 38, William committed himself to study this part of God's creation. He began drawing pictures of snowflakes at age 15. And he wrote, that, "with profound humility, we acknowledge that the Great Designer is incomparable and unapproachable in the infinite. . .beauty of His works." By the time he was 17 years old, Wilson had saved enough money to buy a camera. After a year of diligent labors, he finally had his first photograph of a snowflake. For the rest of his life, he continued to take pictures of snowflakes. Before he died, he gave his entire collection of photographs to the Smithsonian Museum.

Why are no two snowflakes alike? They look a little bit alike when they start out up in the clouds. But, snowflakes take a long trip down to the ground. For over

Snow near Crested Butte, Colorado, USA

GOD MADE THE WORLD

Snowflake Photos Taken by Wilson Bentley

an hour, snowflakes are tossed and turned by the winds, and one by one they are shaped by the air currents. This is God's workshop in the sky.

Snow blesses us in many ways. In addition to watering the earth, snowflakes can absorb dust particles out of the air and also put nitrogen into the soil to fertilize plants. In winter and spring, snow insulates plants above ground and the roots underneath. Ten inches of snow provides about the same amount of insulation as six inches of fiberglass insulation in a building. Two feet of snow can sometimes provide as much as a 40° difference. So, if the air is at -10°F, it might be 30°F (under the snow or a few inches under the ground) at ground level. Many little critters will stay protected under a blanket of snow all winter long.

We are also blessed by a gradual melting of the snow throughout the winter and spring months. Rain will run off the ground quickly, but melting snow slowly seeps into the earth. This prevents flooding, and provides continual watering of new seedlings. Before long, fields will be lush and green with new life covering the earth again. What a tremendous system of irrigation our Creator God has made for His world!

Above all, the Lord uses snow to display His beauty and glory!

CHAPTER 8: GOD MADE DIRT AND WATER

The shining white color reflects all the colors of the light spectrum. A pure white blanket covers the fields and trees with a magical beauty, unmatched by anything else in the world. It is the color of purity. When we are washed clean of our dirty sins, by the blood of Christ He really does make us as white as snow!

"Come now, and let us reason together,"
Says the LORD,
"Though your sins are like scarlet,
They shall be as white as snow;
Though they are red like crimson,
They shall be as wool."
(Isaiah 1:18)

Dirt

God also made dirt. Kids love dirt. Mixed in water, dirt can make great mud pies. But the most important use of dirt is for growing plants.

The Lord knew that all of His animals and men, women, and children would need food. That's the main reason He created dirt or **soil**. Some plants can grow in water, but most plants and trees need dirt to grow. Soil is made up of organic materials, minerals, water, gases, and microorganisms.

Our Father and Creator has provided **organic material** as a very important part of good soil for growing things. Organic material is anything that was once alive—plant or animal. This material decomposes and becomes good nutrition for plants. Soil contains between 1% and 20% organic material, and usually the more of it the better for growing. The best soils are found in Illinois, Minnesota, and the grasslands of Russia and Mongolia. They contain at least 7% of this

GOD MADE THE WORLD

Playing in the Dirt

organic material. These lands are covered with what is called **Mollisol soil**.

The main component in soil is **minerals**, making up about 45% - 49% of its volume. Minerals are necessary for plant growth and health, and they provide good nutrition especially for those who eat the plants. When plants are watered, it's the water that carries the minerals from the soil up into the plant. All of this was planned out by our wonderful Creator.

Oh, that men would give thanks to the LORD for His goodness,
And for His wonderful works to the children of men!
For He satisfies the longing soul,
And fills the hungry soul with goodness.
(Psalm 107:8-9)

Different plants need different kinds of minerals and soils. Thus, farmers and agricultural scientists study plants and soils to make sure the plants are getting what they need. Just as humans need vitamins and minerals, plants need minerals for their health.

The most basic minerals which all plants need are: nitrogen (N), phosphorus (P), potassium (K), magnesium (Mg), sulfur

(S), and calcium (Ca). Thankfully, the Lord included these minerals in most soil types.

Mineral	What it Does
Nitrogen	Gives plants their "green" and helps them produce chlorophyll
Phosphorus	Helps with healthy roots and flower growth
Potassium	Helps plants to grow, to retain water, and to suppress unhelpful insects
Magnesium	Gives plants their "green"
Sulfur	Helps plants resist disease and make seeds.
Calcium	Helps plant cell walls be healthy and strong

Plants also need small amounts of boron (B), chlorine (Cl), manganese (Mn), iron (Fe), zinc (Zn), copper (Cu), molybdenum (Mo), and nickel (Ni).

Microorganisms Crawling Around in the Soil

Soil is made up of more than organic material (decomposed dead things) and minerals. There are also live microbes crawling around in the soil, although you would need a microscope to see them.

These microbes include bacteria, actinomycetes, fungi, protozoa and nematodes. The following table shows what these things do for the soil and plants.

Microorganisms	
Bacteria	Break down nutrients in soil for plant roots
Actinomycetes	Slow down the growth of harmful bacteria; sometimes harmful
Fungi	Help the water and nutrients get into the roots
Protozoa	Eat the bacteria, and then releases nutrients from the bacteria into the soil
Nematodes	Sometimes harmful; some will eat other bad Nematodes and secrete nutrients for the plants

God Made a Variety of Fertile Soils

"But he who received seed on the good ground is he who hears the word and understands it, who indeed bears fruit and

GOD MADE THE WORLD

produces: some a hundredfold, some sixty, some thirty." (Matthew 13:23)

When the seed of the Word of God is sown in the heart, first the soil of the heart must be well prepared to hear it. God provides the good soil. He renews the heart, and prepares us to hear the Word and receive it.

Not all soils are good for growing things. **Fertile soil** is a wonderful blessing from God, as this is soil that yields lots of food at harvest time. Actually, crop yields have improved over the last 50 years. In 1950, the average grain yield was about 880 pounds per acre—or about $440 per acre (for wheat). But, lately that has improved to 2,400 pounds per acre—or about $1,200 per acre. Western European farmers can yield as much as 6,400 pounds per acre—or about $3,200 per acre for wheat. Japanese farmers can yield 4,000 pounds per acre of rice.[1] The average Japanese eats about 100 pounds of rice a year. That means one acre can provide enough rice to feed forty people per year. So, you can see that God has provided a fruitful earth.

CHAPTER 8: GOD MADE DIRT AND WATER

God, in His abundant goodness, has given us as many as 300,000 different kinds of plants we can eat. Amazingly, He did not limit us on variety! However, there are only about 150 plants commonly consumed by people around the world. The most common plant foods include sugar, corn, potatoes, wheat, barley, oats, rice, soybeans, tomato, sorghum, banana, cottonseed, yams, peanuts, millets, sunflower seeds, rye, palm oil, beans, peas, chickpeas, coffee, tea, lentils, cocoa, citrus, sesame, oranges, apples, and assorted nuts. These different fruit-bearing trees and plants will grow in different kinds of soils.

The Lord also made a variety of soils for growing different kinds of plants. These are sandy soil, clay soil, peaty soil, chalky soil, loamy soil, and silty soil. Some soil allows for more air to filtrate through and this can be helpful for root growth. Some soil collects and holds water better than others, and some soils are more dense than others. Soil can also vary in acidity, with varying levels of acid content required for different plants.

Sandy soil is sandy and gritty. Water drains through it easily. It warms up quickly and tends to be acidic. It's easy to cultivate. Minerals wash away easily out of this soil, so you need to add fertilizers, kelp meal, and glacial rock dust. Adding organic material will help it to hold water.

Clay soil will lump together and get sticky when wet. It gets almost rock-hard when it's dry, and can be hard to dig into. Unlike sandy soil, there isn't much air space in this soil. It takes a long time to warm clay soil up in the spring time. However, this soil can be rich in nutrients for plants. Adding organic material will help with the clumping and improve drainage.

Silty soil is made up of fine particles.

247

GOD MADE THE WORLD

It feels soft and light. It can be smushed together easily. Unlike clay soil, silty soil can be washed away by the rain. You can give it a little more structure by adding organic material (compost) into the mix.

Peaty soil is high in organic material. This is usually soil that rotted in a marsh or swamp. It is typically dark and has a spongy feel. Usually, this soil is more acidic and may benefit with the addition of glacial rock dust. Like sandy soil, peaty soil will heat up quickly in the spring time.

It also holds its water well, although it has less nutrients than clay soil. Although rare, this can be great garden soil if you can find it.

Chalky soil has a lot of lime or calcium carbonate in it. This makes it more alkaline (less acidic). It drains well, so you might have to add organic material for improved water retention.

Loamy soil is a mixture of sand, clay, silt, and organic material. This is an ideal soil for gardening, shrubs, and lawns. It

Sandy Soil

Soil Layers

Soil Layers diagram:
- Organic Matter
- Surface Soil
- Subsoil
- Parent Rock
- Bedrock

tends to be acidic, and you will need to add nutrients from time to time.

Using God's Soils—Gardening and Farming

Then the LORD God took the man and put him in the garden of Eden to tend and keep it. (Genesis 2:15)

From the beginning, God wanted man to take good care of the plants and the trees in the garden. He doesn't want all the plants to grow wild. And, He will not provide us with all of our food without some work on man's part. Adam was supposed to tend the Garden of Eden.

People would never be able to live on wild plants. Without farming and gardens, most of us would starve to death.

After man fell into sin, God cursed the ground. He made it harder to bring fruits out of the earth by making sure that there would be more weeds and thistles all over the world.

Then to Adam [the LORD God] said, "Because you have heeded the voice of your wife, and have eaten from the tree of which I commanded you, saying, 'You shall not eat of it':
Cursed is the ground for your sake;
In toil you shall eat of it
All the days of your life.
Both thorns and thistles it shall bring forth for you,
And you shall eat the herb of the field.
In the sweat of your face you shall eat bread
Till you return to the ground,
For out of it you were taken;
For dust you are,
And to dust you shall return."
(Genesis 3:17-19)

GOD MADE THE WORLD

Soil Type	Plants that Grow Well in this Soil
Sandy Soil	Fig trees, tulips, hibiscus, carrots, potatoes, lettuce, strawberries, peppers, corn, tomatoes, squash
Clay Soil	Summer vegetables, fruit trees, shrubs, ornamental trees (flowering quince)
Silty Soil	Vegetables and fruits, birch, dogwood, willow, cypress trees, shrubs, climbers, grass, perennials
Peaty Soil	Legumes, root crops (carrots, turnips), cabbage, spinach, heather, witch hazel, rhododendron
Chalky Soil	Trees, shrubs (lilac), Sweet corn, spinach, beets, and cabbage
Loamy Soil	Perennials, climbers, bamboo, vegetables, berries

Now because of sin it takes a lot of hard work to bring food out of the earth. Agricultural (farming) scientists are called to study plants, and better understand soils, pests, and weeds to help the soil bring forth useful foods.

Farming and Gardening

"I will bring back the captives of My people Israel;
They shall build the waste cities and inhabit them;
They shall plant vineyards and drink wine from them;
They shall also make gardens and eat fruit from them.
I will plant them in their land,
And no longer shall they be pulled up
From the land I have given them,"
Says the LORD your God. (Amos 9:14-15)

The farmer (or gardener) has six main jobs to do.

1. First, he must choose the sort of plants he will grow. This is based on the soil type available, weather patterns, the sort of pests he will have to deal with, the needs of his family,

and the market of people who want to buy the food. He needs to think about how much sunlight is available as well. Most vegetables and flowers need six to eight hours of sunlight a day.

2. The farmer must prepare his land for planting. He turns up the soil, adds fertilizer and nutrients, removes sod or weeds, and pulls out large stones and rocks. Most gardens need added organic matter—compost, grass clippings, and decayed leaves. If the soil is loose enough, it may not need to be worked. Gardeners will either till the soil with a machine or dig it up with a shovel. Too much tilling can damage the soil, and disturb the earthworms and other helpful organisms. When digging into the soil, the gardener will turn the top 8-12 inches of soil gently. All the while, he mixes the organic material into it.

3. The farmer plants his seeds or seedlings. Small plants already nurtured in a greenhouse are called seedlings. The gardener is especially careful to space his plants far enough apart so they don't compete for nutrients and water.

4. Unless the seeds can sprout on their own with natural moisture, the farmer will have to irrigate his garden. Young plants often need daily watering, and this can taper off as the plants get larger and

GOD MADE THE WORLD

stronger. Added mulch will keep moisture in and weeds out. If the soil about three inches under the surface feels dry, the plants likely need more water. Remember that clay soil holds moisture better than sandy soil. Also, sun and wind can dry out the soil quickly. Water should be sprinkled over the garden, not poured, to give it a chance to soak into the dirt.

5. The farmer will have to continue monitoring his crop growth, eliminating weeds, fertilizing, and watching out for pests.

6. Finally, the farmer harvests his crops, and gives God thanks for the increase.

Let the peoples praise You, O God;
Let all the peoples praise You.
Then the earth shall yield her increase;

CHAPTER 8: GOD MADE DIRT AND WATER

God, our own God, shall bless us.
God shall bless us,
And all the ends of the earth shall fear Him. (Psalm 67:5-7)

God Promises a Good Harvest

"They shall build houses and inhabit them;
They shall plant vineyards and eat their fruit.
They shall not build and another inhabit;
They shall not plant and another eat;
For as the days of a tree, so shall be the days of My people,
And My elect shall long enjoy the work of their hands." (Isaiah 65:21-22)

Throughout God's Word we find the encouragement to plant seed. When Jesus comes, God promises fruitfulness in the earth—both spiritual and physical. In Isaiah 65, the Lord challenges us to plant vineyards and eat the fruit of them. Of course, the most important seed we will plant is the Word of God. This seed brings forth good fruit all around us. But let us also plant gardens and pray for good harvest. Let us look to God to

George Washington Carver

Carver (1865-1943) was one of America's greatest agricultural scientists. He rescued farmers in the American South when boll weevils were destroying their cotton crops in the early 1900s. Carver also expanded Southern farming, to include sweet potatoes and peanuts, and he developed many non-food uses for these plants. His discoveries included hundreds of different kinds of cosmetics, household products, paints, glues, foods, beverages, and medicines. Dr. Carver was very humble when it came to his discoveries. He explained that, "I didn't make these discoveries. God has only worked through me to reveal to His children some of His wonderful providences."[2]

George Washington Carver was known to be a very loving and kind man, often giving away his money to those in need. Through his generosity and his inventions, he sought to be a blessing to others. And, by God's mercies, his inventions have blessed billions of people around the world.

Rice Fields in Vietnam

fulfill His promise in Isaiah 65.

Even when the people of God go through times of trial, the Lord still tells them to plant gardens. During hard times, we must still look to God for the provision of a blessed harvest. ▪

> Thus says the LORD of hosts, the God of Israel, to all who were carried away captive, whom I have caused to be carried away from Jerusalem to Babylon. Build houses and dwell in them; plant gardens and eat their fruit. Take wives and beget sons and daughters; and take wives for your sons and give your daughters to husbands, so that they may bear sons and daughters—that you may be increased there, and not diminished. (Jeremiah 29:4-5)

Pray

- Praise God for the provision of water, soil, sun, and air. He has given us everything we need to grow food for animals and for man. Thank Him for the grains, the fruit, and the vegetables that grow every year for our food. Be amazed at how water and minerals are carried up into tree trunks, branches, and leaves.
- Praise Him for such mysterious forces that can oppose gravity. Praise God for the beauty of snowflakes, and the blessing of snow on the earth.

Sing

Can a Little Child Like Me?

Can a little child like me
Thank the Father fittingly?
Yes, oh yes! be good and true,
Faithful, kind, in all you do;
Love the Lord, and do your part;
Learn to say with all your heart,

Refrain:
Father, we thank Thee,
Father, we thank Thee,
Father in Heaven, we thank Thee.

For the fruit upon the tree
For the birds that sing of Thee,
For the earth in beauty dressed,
Father, mother, and the rest,
For Thy precious, loving care,
For Thy bounty everywhere,

For the sunshine warm and bright,
For the day and for the night,
For the lessons of our youth—
Honor, gratitude, and truth,
For the love that met us here,
For the home and for the cheer,

For our comrades and our plays,
And our happy holidays,
For the joyful work and true
That a little child might do,
For our lives but just begun,
For the great gift of Thy Son,

If you do not know this song, you may listen to a version of the hymn on the Internet, with supervision, and sing along with it.

Do

Take the lessons learned in this chapter, and apply them in real life by creating a small garden.

Create a small garden or grow something in a small planter box. A fun little garden can be grown in just a square yard (meter). Here are the steps to take:

1. **Purchase seeds, seedlings, or sets. Seeds:** Lettuce, radish, bush bean, and spinach seeds can be sown directly into the garden. These plants work well in small gardens since they don't take up much space. It is not necessary to plant in rows in a tiny garden. The plants can just be spaced according to the seed packet directions. Pumpkins, zucchini, peas, and corn would work better in a larger garden where they can spread or grow tall. **Seedlings:** Tomato and pepper seeds can be planted indoors in a sunny window six to eight weeks before the last frost outside. Planting these small plants (seedlings) into the garden, instead of seeds, gives them a head start and more time to produce. Seedlings can also be purchased. Choose a cherry tomato variety for an earlier crop and a smaller plant. **Sets:** Onions and garlic are often planted from sets (immature bulbs grown the previous season). When planted they finish growing and yield a much larger bulb than if started from seed.
2. **Choose a site.** Gardens must get at least six hours of sunshine a day to produce. They also need to be watered daily at first, so make sure there is access to water.
3. **Prepare the soil.** Turn over the soil 8 - 12 inches (20 - 30 cm) deep, with a shovel. Spread well-rotted organic matter two inches (5 cm) deep over the garden and work in with a shovel or hoe. Don't work the soil if it is muddy or it will become compacted. Wait until it is just damp. Try not to compress the soil by stepping on it.
4. **Plant.** Seeds must be kept moist until they sprout. Sprinkle gently to avoid washing the seeds away. Seedlings are best transplanted on a cool day or late in the day to give them a chance to adjust before dealing with heat. (It is also a good idea to allow the plants adjust to the outside world during the daytime for a week or two before transplanting; this is often referred to as hardening off.)
5. **Patrol for pests.** Pull weeds or scrape them off at ground level if there is a risk of disturbing the roots of your crop. Pluck off large insect pests, slugs, or snails. Small

insects, like aphids, can be sprayed off with a stream of water. Leave beneficial insects, like ladybugs, praying mantises, and wasps, so they will eat harmful pests. Larger pests, like rabbits and voles, may need fence barriers or trapping.
6. **Harvest with thankfulness.** Lettuce can be cut off a little above the ground. The remaining stump will grow small lettuce leaves for another harvest. Lettuce will get bitter after it bolts (grows into a tall spike) so have lots of salad early. Harvest radishes before they get woody, and beans before their seeds swell in the pods. Tomatoes can stay on the plant a while after turning red (without spoiling).

Remember to put your tools away so they don't rust and you can find them next time you need them.

Watch

To watch the recommended videos for this chapter, go to **generations.org/ GodMadeTheWorld** and scroll down until you find the video links for Chapter 8. Our editors have been careful to avoid films with references to evolution. However, we would still encourage parents or teachers to provide oversight for all internet usage. These videos may not give God the glory for His amazing creative work, so the student and parent/teacher should respond to these insights with prayer and praise.

Storms in Ireland

Chapter 9

GOD GIVES US ATMOSPHERE AND WEATHER

> The LORD is slow to anger and great in power,
> And will not at all acquit the wicked.
> The LORD has His way in the whirlwind and in the storm,
> And the clouds are the dust of His feet.
> He rebukes the sea and makes it dry,
> And dries up all the rivers.
> Bashan and Carmel wither,
> And the flower of Lebanon wilts. (Nahum 1:3-4)

God planned that there would always be changes going on in our world. This is especially so with the weather. In the Gospels, we learn about how Jesus' disciples were in a boat on the Sea of Galilee when the winds and the waves picked up. This was a test of their faith. The changes of weather can be hard on all of us. But even more so, violent storms can bring us to fear. Through it all, let us remember that God is in charge of the weather. All of these things come from His hand. He brings the storms and He brings the calm.

The Sky

He made the Pleiades and Orion;
He turns the shadow of death into morning

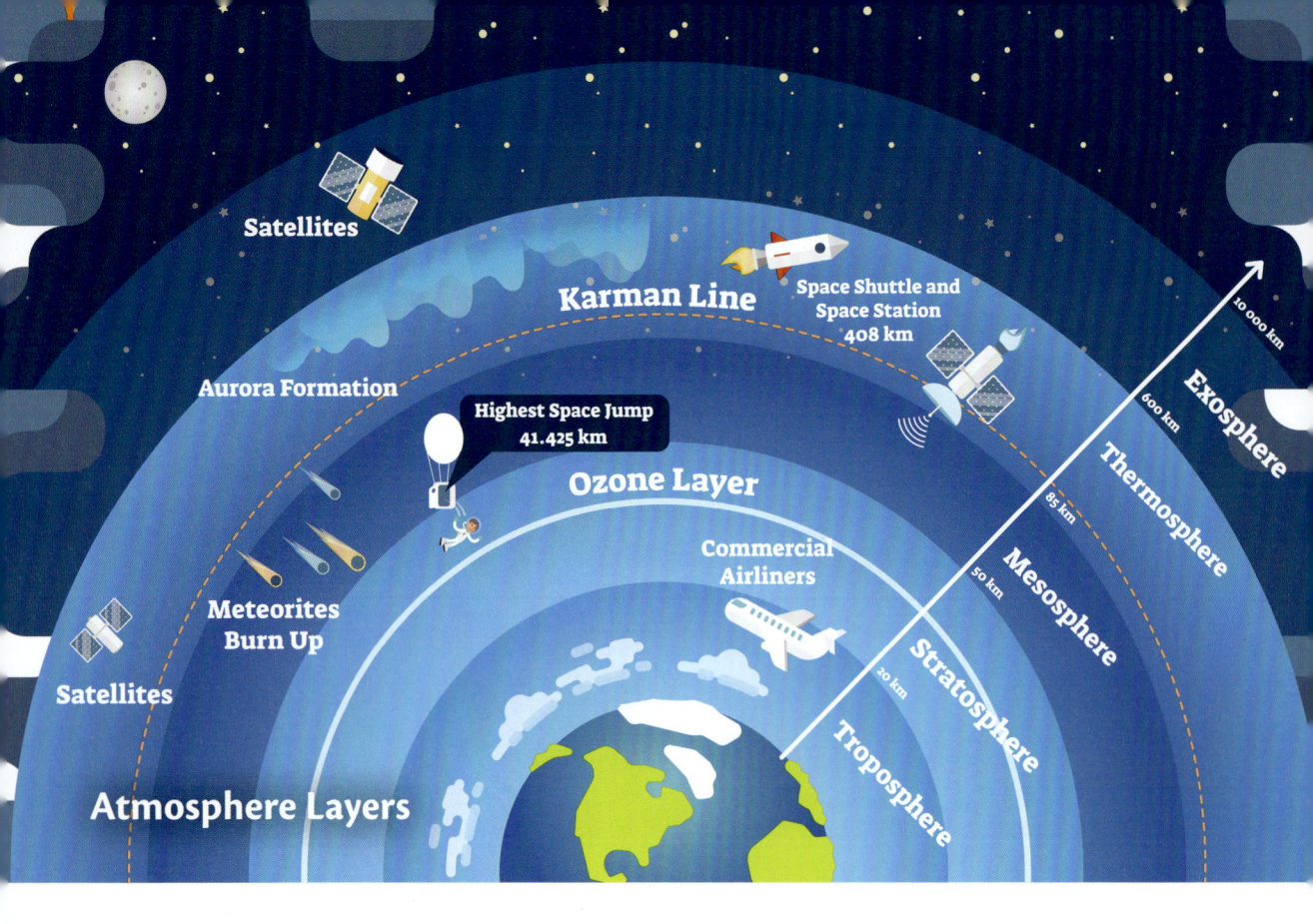

Atmosphere Layers

- Satellites
- Karman Line
- Space Shuttle and Space Station 408 km
- Aurora Formation
- Highest Space Jump 41.425 km
- Ozone Layer
- Meteorites Burn Up
- Commercial Airliners
- Satellites
- Exosphere 10 000 km
- Thermosphere 600 km
- Mesosphere 85 km
- Stratosphere 50 km
- Troposphere 20 km

*And makes the day dark as night;
He calls for the waters of the sea
And pours them out on the face of the earth;
The LORD is His name. (Amos 5:8)*

Our Creator God made the world with an atmosphere of air all around it. This is what we call a canopy of gases that surround the earth. The atmosphere is made up of five layers: the troposphere, stratosphere, mesosphere, thermosphere, and the exosphere.

The troposphere is the layer closest to the earth, where we live and breathe. All of our weather patterns, the clouds, and rain develop in the troposphere. The troposphere contains about half the air in our atmosphere. If you were to ascend in a balloon, where the large jets fly, the air would become very cold—about -60°C, at the top of the troposphere.

After the troposphere comes the stratosphere, which contains the ozone layer. Whereas oxygen gas is a molecule made up of two atoms, ozone gas is made up of three atoms, designated by the symbol O_3. This layer of ozone protects the earth from the sun's harmful ultraviolet radiation. When these rays hit the ozone molecules, the energy turns into harmless heat. As your balloon continues to ascend through the stratosphere, the temperatures would get warmer, climbing up to about -3°C.

The next layer above the stratosphere is the mesosphere. This is where meteors usually burn out as they approach the earth. Now, the temperatures begin to grow colder as you work your way up through the mesosphere—as low as -130°F (-90°C).

Above the mesosphere is the thermosphere where manmade satellites revolve around the earth. God has designed the thermosphere to absorb much of the dangerous X-Ray and Ultraviolet radiation, so the temperature can get very hot—between 932°F (500°C) and 3,632°F (2,000°C). The beautiful **northern and southern lights** appear in the thermosphere. These lights are created by electrons mixing with gases blown about by solar winds.

The exosphere is the final layer of our atmosphere before you get to space. It is separate from space because there is still enough gravity to keep some gas molecules from escaping. The Hubble Space Telescope and satellites orbit around the

Clouds in the Troposphere

earth in the thermosphere and exosphere.

When you look up into the sky, you see clouds, meteors, the moon, the sun, and the stars. By the human eye, it's hard to know exactly how far away these objects are from where you stand. The following table will give you some idea of the distance of far-off objects from the earth.

Object	Altitude from Sea Level
Summit of tallest mountain (Everest)	5.5 miles
Airliners' highest flight path	7 miles
Troposphere ends	12 miles
Meteors burning up in mesosphere	45 miles
Lowest satellites	310 miles
Most communication satellites	20,000 miles
Moon	240,000 miles
Sun	94,000,000 miles

Air Pressure

The atmosphere is full of gas molecules that add up to a lot of mass. While solids and liquids are heavier, gases make for less weight in a given space. All the mass in the atmosphere rests on the earth by gravity, and this produces a pressure called **atmospheric pressure**. Although you can't feel it, the atmosphere is pushing in on you all the time. Atmospheric pressure is about 14.7 pounds per square inch (psi) or 100 kilo-pascals. That's like a 14.7-pound weight sitting on every square inch of your body. You can't feel it because your body is used to it. The ultimate reason you can't feel it is because the air inside your body is pushing out at the same level of pressure as the air is pushing in on your body from the outside.

The higher you go in altitude, the lower the atmospheric pressure. For example, the atmospheric pressure in Denver is about 12 psi (83 kPa). If a bag of chips was manufactured and sealed in Los Angeles (at sea level), it would blow up like a balloon if you were to bring it up into the Rocky Mountains to Denver, Colorado. That is because the pressure inside the bag is about 14.7 psi, while the pressure outside the bag is 12 psi. When

CHAPTER 9: GOD GIVES US ATMOSPHERE AND WEATHER

A Bottle at 14,000 Feet (4,300 m), 9,000 Feet (2,700 m) and 1,000 Feet (300 m)

you open the bag in Denver, you will hear a little pop. That comes from the release of pressure from inside the bag.

Opening yogurt at higher altitudes can be messy too. If yogurt is packaged at an atmospheric pressure of 13.5 psi, and it is opened at a higher altitude where the atmosphere is 11 psi, it will spurt out on you. When you open yogurt at high altitudes, peel the lid with the opening facing away from you—and preferably, not facing your brother or sister.

If you seal an empty water bottle at the top of a 14,000-foot high mountain what do you think it will look like at the bottom of the mountain?

Suppose your family took a trip from sea level to the mountains. Before you left, you filled your car tires to 30 psi. If you drove to the top of a 10,000-foot mountain pass, the atmospheric pressure outside would change from 14.7 psi to 10.1 psi. Would your tires lose or gain pressure? Because the pressure pushing on the tires from the outside is less, the pressure inside would actually increase by 4.4 psi.

If you measure the atmospheric pressure over and over again in one place, you will find that it will change.

263

GOD MADE THE WORLD

Car in Mountain Pass, North Cascades National Park, Washington, USA

The highest atmospheric pressure ever recorded was 15.75 psi at sea level. The lowest atmospheric pressure ever recorded at sea level was 12.61 psi.

Altitude is the main thing that determines atmospheric pressure. Also, changes in temperature can affect atmospheric pressure. Changes in atmospheric pressure are monitored to predict changes in the weather.

There is a difference between **atmospheric pressure** and **air pressure**. Atmospheric pressure is the pressure from the air in the atmosphere pressing down on all of us. But, the term air pressure is used for the pressure within a closed container (like a bicycle tire). Atmospheric pressure changes very little from summer to winter. However, the pressure in your bike tires or car tires can change when the temperature changes from hot to cold. Temperature of the air (gas) has a direct relationship to gas pressure. The higher the temperature, the higher the pressure. A good rule of

CHAPTER 9: GOD GIVES US ATMOSPHERE AND WEATHER

thumb would be an increase of about 1 psi for every 10 °F. Suppose you pumped up your bike tire to 24 psi on a cold night in March (when the temperature was 20 °F). Later in the summer, as the temperature rose to 80 °F, the air pressure in the same tire would be about 30 psi. That assumes no air leaked out in the meantime. When the temperature rose 60°, the bike tire would be about 6 psi higher.

The barometer is an instrument used to measure the amount of atmospheric pressure. This equipment is very important for **meteorologists** and pilots.

Weather

God understands [Wisdom's] way,
And He knows its place.
For He looks to the ends of the earth,
And sees under the whole heavens,
To establish a weight for the wind,
And apportion the waters by measure.
When He made a law for the rain,
And a path for the thunderbolt,
Then He saw wisdom and declared it;
He prepared it, indeed, He searched it out.
And to man He said,
"Behold, the fear of the Lord,
that is wisdom,
And to depart from evil is understanding."
(Job 28:23-28)

Weather is a big part of our lives. We go on picnics when the weather is sunny, and we stay indoors during storms. Weather can be deadly. Every year, 20,000 - 100,000 people die from weather-caused tragedies like heat, storms, tornadoes, hurricanes, lightning, and wildfires. Forest fires are usually caused by excessively hot, dry weather.

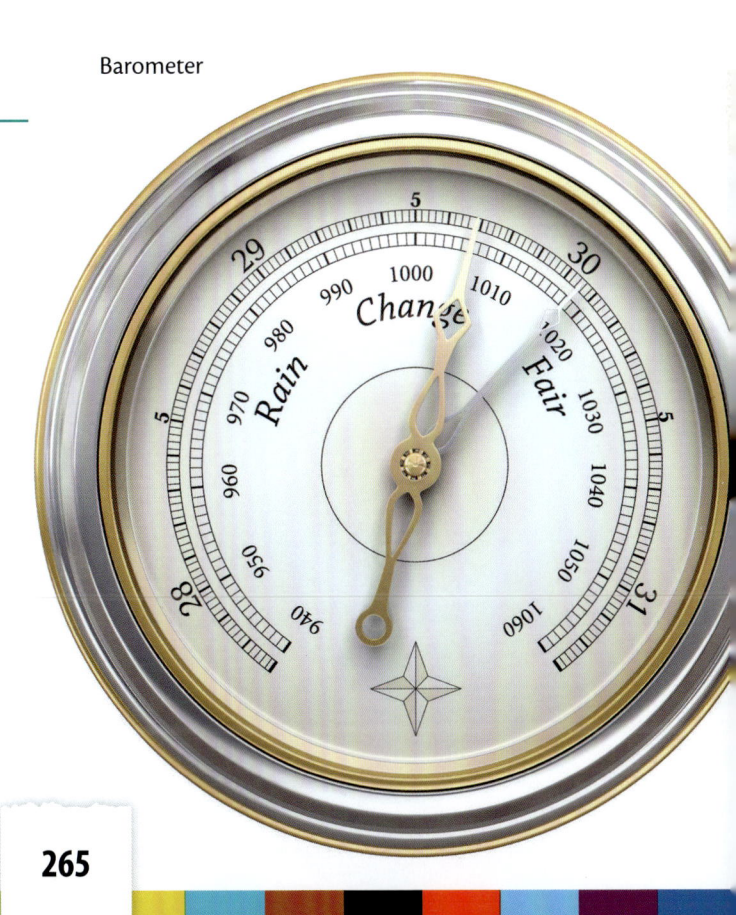

Barometer

GOD MADE THE WORLD

Because weather changes so much, it is good to know the things that cause these changes. The first thing you need to know is that God made the sun to have the biggest influence on our weather. The heat energy from this giant burning fire is more important than everything else. We can't control sunspots, solar flares, or any of the changes on the surface of the sun that account for the weather patterns. But we know that God does control all of it.

Because earth is tilted, the sun heats some parts of the earth more than others at different times of the year. This is the major cause for different weather conditions throughout the year.

The sun does not warm up the air above the earth. Rather, the sun warms the surface of the earth, and that is what warms up the air around us.

Wind

Cooler areas usually have higher air pressure. That's because the air molecules huddle together closer. Because heat rises, the air that is warmed up by the earth rises and this creates a low pressure. The earth is cooler in some areas and warmer in oth-

Tornado

Lightning

ers. So, there will be higher atmospheric pressure in some areas than others. With these pressure differences, air will begin to move from the higher pressure to the lower pressure areas where the earth was warmer. Wherever air is moving, you can feel it blowing, and that is called **wind**.

Air masses are huge bodies of air, hundreds of miles in width and length. When these large air masses move from one part of the country to another, the winds will blow for days at a time. When God brings two air masses together in a collision, it's called a **front**. When a cold air mass plows under a warm mass, it's a **cold front**. When a warm mass climbs over a cold mass, it's called a **warm front**. The warm front usually produces fog and a light rain. After a few days, the land area under this warm front will warm up. Cold fronts usually produce thunderstorms and rain, and sometimes hailstorms.

A **stationary front** develops when the warm and cold fronts press against each other with an equal force. Often the stationary front sits there for several days, producing clouds and rain.

Heavy Wind During a Hurricane

Sea and Land Breezes

Coastal areas and islands will often get sea breezes. The land heats up during the day when the sun shines on it. As warmer air over the land rises (often at night), the cool air over the cooler sea waters will be drawn into land. This is what creates the sea breeze blowing inland. If the sea water is warmer than the land, which can happen in the early morning hours, the breezes blow from the land towards the sea.

Where Does All This Rain Come From?

"But I say to you, love your enemies, bless those who curse you, do good to those who hate you, and pray for those who spitefully use you and persecute you, that you may be sons of your Father in heaven; for He makes His sun rise on the evil and on the good, and sends rain on the just and on the unjust." (Matthew 5:44-45)

Rain is one of the most wonderful blessings in all the universe. The continual cycle whereby God gives the blessing of rain, year in and year out, is truly amazing. At any point in time, there are about 3,100 cubic miles (13,400 cubic kilometers) of water in the sky. That's more water than what is contained in all the rivers in the world. Humans use about 4,000 cubic kilometers of water in a year for drinking water, baths, and irrigating farms. In total, God sends about 510,000 cubic kilometers of water upon the earth in a year. That's about 150 times the amount needed for the 7 billion people on the earth. How merciful God is to the world, that He would send that much water for man and animal!

How does all this water get up in the

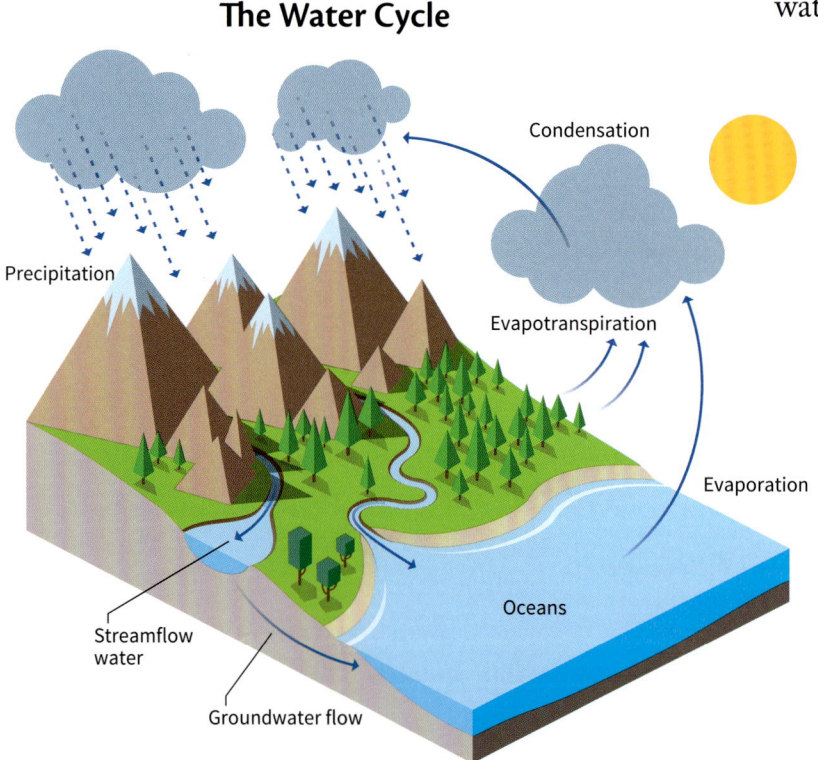

The Water Cycle

- Precipitation
- Condensation
- Evapotranspiration
- Evaporation
- Oceans
- Streamflow water
- Groundwater flow

CHAPTER 9: GOD GIVES US ATMOSPHERE AND WEATHER

Rain in Kerala, India

clouds year after year? What wondrous design has the Lord ordained to bring all this rain on the earth? For 6,000 years, God has provided precious water for plants, animals, and man.

Clouds form by evaporation and condensation. **Evaporation** is the process where water turns into vapor, when the temperature and/or pressure increases on the earth's surface. **Condensation** occurs when water vapor turns back into water droplets, as the temperature and/or pressure drops on the earth's surface. This is how God makes clouds and rain. Condensation is the opposite of evaporation. Water is condensing when your glasses fog up when you walk out of a cold room to the outdoors on a hot, humid day. The same thing is happening when you see water drops forming on your glass of cold soda.

Evaporated water from the ocean and land floats around in the air (in gas form). It has to turn back into liquid water before it falls back onto the earth. But how does this happen? First, the sun warms up certain parts of the earth, and a blob of warm air forms on the ground. Because hot air always rises, this blob of warm air pulls away from the earth and begins to float upward. This blob of warm air is called a **thermal**. Birds and hang gliders like to find these thermals, because they help them to soar higher and higher in the sky.

Now, the higher the thermal rises the more it will spread out and cool off. As

Nimbostratus Clouds

Types of Clouds

Cloud Type	Looks Like	Other Info
Cirrus	Thin and wispy	Made of ice crystals
Cirrocumulus	Fluffy and arranged like fish scales	Does not rain
Cirrostratus	Blanket-like, sun and moon shines through	Made of ice crystals
Altocumulus	Fluffy, appear in groups	Light and moderate rain
Altostratus	Grey, smooth, spreads over many miles	Light rain or snow
Nimbostratus	Thick, dark, heavy clouds	Heavy rain
Stratus	Thin layer spreading over large area	Fog, light showers
Cumulus	Fluffy cotton ball, well defined edges	Only occasional light shower

the thermal air mass cools, the moisture in the air will begin to condense into tiny droplets of water. As long as warm air continues to rise, the clouds will stay up in the sky. You can watch these clouds blow around by the winds. The average water molecule spends about ten days blowing around in the air before it comes back down in rain. Small particles of dust and dust clouds can also float around in the atmosphere before they settle back down.

The particles of water floating in the

clouds are very small. They are so small, it takes about one million of these micro-droplets to form one rain drop. As clouds cool down, the micro-droplets combine with each other. Finally, when the big raindrops become too heavy to stay in the sky, they fall down on the earth as rain. This is how God makes rain. What a convenient way to distribute fresh water everywhere, without using pumps, hoses, pipes and aqueducts!

There are different kinds of clouds in the sky. Clouds help meteorologists figure out what the weather is going to be like today or tomorrow.

Clouds can be divided into three categories, based on how high they are in the sky. **High-level clouds** include cirrocumulus, cirrus, and cirrostratus (5-13 km high). **Mid-level clouds** include altocumulus, altostratus, and nimbostratus (2-7 km high). And, **low-level clouds** include stratus, cumulus, cumulonimbus, and stratocumulus (0-2 km high).

"Behold, God is great, and we do not know Him;
Nor can the number of His years be discovered.
For He draws up drops of water,
Which distill as rain from the mist,
Which the clouds drop down

And pour abundantly on man.
Indeed, can anyone understand the spreading of clouds,
The thunder from His canopy." (Job 36:26-29)

Hail

Praise the LORD from the earth,
You great sea creatures and all the depths;
Fire and hail, snow and clouds;
Stormy wind, fulfilling His word. . .
(Psalm 148:7-8)

Hail forms as warm summer air pushes moisture higher and higher in the sky. Miles up above the earth, temperatures drop and the moisture freezes into icy pellets. These pellets fall back down, but then the warm air pushes them back up to where more layers of ice form on the pellets. The process continues until the hail is large enough that the warm air cannot hold the hailstones up anymore, and they fall to the ground.

The largest hailstone ever recorded was eight inches in diameter, found in Vivian, South Dakota. Three people have been killed by hailstones in the United States. In China, 25 people were killed in

GOD MADE THE WORLD

one hail storm which occurred on July 19, 2002. God used hail in a powerful way to help His people win a battle against the enemy in Joshua 10.

Hailstorm in Tennessee, USA

So Joshua ascended from Gilgal, he and all the people of war with him, and all the mighty men of valor. And the LORD said to Joshua, "Do not fear them, for I have delivered them into your hand; not a man of them shall stand before you." Joshua therefore came upon them suddenly, having marched all night from Gilgal. So the LORD routed them before Israel, killed them with a great slaughter at Gibeon, chased them along the road that goes to Beth Horon, and struck them down as far as Azekah and Makkedah. And it happened, as they fled before Israel and were on the descent of Beth Horon, that the LORD cast down large hailstones from heaven on them as far as Azekah, and they died. There were more who died from the hailstones than the children of Israel killed with the sword. (Joshua 10:7-11)

Humidity, Dew and Frost

The amount of moisture hanging

Dew on Grass

out in the air is called **humidity**. Hot air can hold a lot more moisture than cold air. When you get out of bed in the morning and look outside, you may see dew or frost in the yard. Because the air has cooled overnight, it cannot hold as much moisture as warm air can. So in the cool mornings, the moisture which was in the air (as a gas) settles on the surface of the earth as a liquid—called **dew**. When the temperature on the ground is less than 32 °F (0 °C), the moisture from the air freezes on the ground and on trees and bushes—called **frost**.

Severe Weather

Bless the LORD, O my soul!
O LORD my God, You are very great:
You are clothed with honor and majesty,
Who cover Yourself with light as with a garment,
Who stretch out the heavens like a curtain.
He lays the beams of His upper chambers in the waters,
Who makes the clouds His chariot,
Who walks on the wings of the wind. . .
(Psalm 104:1-3)

GOD MADE THE WORLD

Most days are fairly calm and peaceful, when it comes to our weather, but sometimes the Lord brings mighty storms on the earth. This shows His power. He doesn't want us to be afraid of the thunder and lightning, the floods and fires. Rather, He sends these powerful storms so that we will stand in awe of His power, and fear and worship Him.

We learned earlier that lightning is a display of God's powerful electricity. Air movement in the clouds creates a gigantic show of static electricity. This is the same energy you see when you take off your sweater, and sparks fly. But God's lightning is much more impressive, heating up to 54,000 °F (30,000 °C). This intense heat instantly warms all the air around it, and this increases the air pressure—sort of like filling a balloon with a lot of air under high pressure. All that compressed hot air explodes, and that is what causes the loud thunder crack. This all happens in a split second—the flash and the thunder crack.

Since light travels faster than sound, you will see the lightning before you hear the thunder crack. To figure out how far away the lightning struck, count the seconds between the flash and the thunderclap. Multiply the seconds by

Lightning Above Ft. Worth, Texas, USA

1000 feet (300 m). A five second delay would be 5,000 feet or about one mile away (1,500 m).

About one in three thousand people are struck by lightning at least once in a lifetime. Only about 10% of people hit by lightning will die of the strike. A park ranger, Roy Sullivan, was hit by lightning seven times and survived! This is an extremely unlikely occurrence, and would only happen to one person in 10,000,000,000,000,000,000,000,000,000 people. Ultimately, God is in control of where lightening strikes, and only He can preserve our lives.

Blessed be the Lord,
Who daily loads us with benefits,
The God of our salvation! Selah
Our God is the God of salvation;
And to GOD the Lord belong escapes from death. (Psalm 68:19-20)

Hot and Cold Temperatures

Both cold and hot temperatures can be dangerous for humans, but cold conditions kill about 17 times more people than the heat. More people will die of long-term exposure to cold than any other temperature concern.

By God's design the body tries to regulate its temperature to 98.6 °F. People exposed to heat above 91 °F (33 °C) without shade for a long time may wind up with heat exhaustion or heat stroke. That's when the body temperature goes over 104°F. Strenuous activity and not drinking enough fluids on a hot day is what produces heat stroke. Also, people visiting warmer climates who are used to colder climates are more likely to get heat stroke.

People suffering from heat stroke will show some of the following signs:

- Confusion
- Slurred speech
- Irritability
- Seizures
- Coma
- Nausea and vomiting
- Rapid breathing
- Red flushed skin
- Racing heart rate
- Headache

If you think somebody is getting heat stroke, you should get the person out of the sun as soon as possible. Remove excess clothing, and use cool water, spray mists, cold packs and wet cloths to cool them down.

GOD MADE THE WORLD

If you are ever lost in the desert, the most important thing to keep in mind is **dehydration**. To avoid this, stay in the shade or if there isn't shade you might bury yourself in the sand, shading your head with you backpack or jacket. Build a shelter if you can. Don't wander around in the hot sun. Keep your clothes on to avoid sunburn. Sweating is good because that's God's design for cooling you off on hot days. Don't waste your time and energy hunting for food. Most humans could survive for three weeks without food, but not more than a day or two without water.

If possible, you should make a fire to signal your location to rescuers. Cactus water is usually not drinkable, so you should avoid it except as a last resort. Ground temperatures are usually pretty high, so it would be better to sit on something and keep yourself off the ground.

Protecting Yourself in the Cold

Any temperature below freezing is dangerous for humans, if they stay out in the cold unprotected for more than 30 minutes. Wet socks and clothing can be very dangerous because water is thirty

Sahara Desert, Morocco

Snow-Covered Highway in Austria

times more conductive than air, and quickly transfers heat away from the body. Frostbite is of first concern when you are out in the cold, unprotected from the elements. That is where ice crystals form in the skin and restrict blood flow. A person who stays out in 0°F weather, with the wind blowing at 15 mph, would end up with frostbite within 30 minutes if some portion of their body was unprotected.

Another concern with going out into the cold is **hypothermia**, which means "low heat." This is a condition where a person's body temperature drops dangerously low. To avoid this, you should bundle up in warm clothes, and keep your body going on high-carbohydrate snacks.

Treating somebody with hypothermia usually begins by warming the body from the inside out. The victim needs to drink warm liquids, and keep warm clothes on the head and trunk of his body. Warming arms and legs too quickly, without warming the trunk, could cause blood circulation issues.

Snowstorms

Snowstorms can be deadly. The

GOD MADE THE WORLD

most important thing to know about snowstorms is to stay out of them. Don't travel. Stay inside, and don't walk around in blizzards.

If you happen to be stranded in a car in a snowstorm, do not leave the vehicle. Unless there is a building within sight of where you are, the safest thing to do is to stay sheltered in the car.

1. **Run the car enough to keep it comfortably warm.** Use all the clothes and blankets and car mats available to you for warmth.

2. **If you have a cellphone call someone who can help.** Pinpoint your location using your GPS. Inform the rescuers of your condition, your location, and the extent of your resources—gas, food, and water.

3. **Try to make yourself visible** for any rescuers nearby by turning on the car flashers and rigging up a dark colored flag (that can be seen against the white snow). You could honk out SOS, the international signal for "help," using your car horn. That's three long beeps, three short beeps, and three more long beeps.

4. **Monitor your gas situation.** A full tank of gas could keep the vehicle idling and provide you with heat for 24-36 hours depending on the car model.

5. **Regularly clear out the snow gathering around your exhaust pipe,** to avoid the risk of carbon monoxide poisoning in the car. Keep your lights

clear of snow, especially if you hope to be noticed by rescuers.

6. **Keep high-carbohydrate snacks in the car,** if you are traveling (or living) in areas where snowstorms are likely.

If you are caught in a blizzard without shelter, the following is an excellent survival plan:

1. Build a snow shelter. Remember, snow is a very good insulator from the cold. Shape it like an igloo by building a mound and then digging out an inside shelter.

2. You will need to have a small hole (cut in the roof) about two inches across in the roof. Cut another hole in the side for ventilation.

3. For a bed, sleep on branches, leaves, and blankets. Avoid sleeping directly on the snow. Your body heat will melt the snow, and the water will rob you of heat. If you are with someone else, share covers and warmth.

4. Continue to drink water, to keep yourself hydrated. Be sure to let the snow melt before taking it in. Keep your body covered—hat and gloves, at all times.

5. Stay in one place. Walking in deep snow can wear you out quickly, especially if you are lost and you don't know where you are going.

6. After the snowfall stops, spell out the word "HELP" using branches or rocks in a clear area. This will help a search party see your location from a helicopter or airplane.

7. And most importantly, pray for God's protection and a quick rescue.

Hurricanes

While he was still speaking, another [messenger] came and said, "Your sons and daughters were eating and drinking wine in their oldest brother's house, and suddenly a great wind came from across the wilderness and struck the four corners of the house, and it fell on the young people, and they are dead; and I alone have escaped to tell you!" Then Job arose, tore his robe, and shaved his head; and he fell to the ground and worshiped. And he said:
"Naked I came from my mother's womb, And naked shall I return there.
The LORD gave, and the LORD has

**taken away;
Blessed be the name of the LORD."
(Job 1:18-21)**

Powerful storms come by the command of God. All weather events show His sovereign power. How might we respond to the display of God's power? Job's response to a powerful natural disaster in his own life was to fall on his face and worship the Lord. He realized that God was in control of all of this, and he knew that worship was the most important part of his life.

The most dangerous weather condition in the world is what is called a **hurricane**, also called a **typhoon** in the Pacific. Hurricanes form near the equator or where ocean waters are warm—80 °F or higher. An enormous amount of water evaporates, with global winds blowing across the oceans helping the process along. The water vapor rises high into the sky, cooling and forming huge clouds. Winds begin to blow in a spiral, circular manner, gathering many of the large clouds into one gigantic cloud. If the spinning winds exceed 74 miles per hour, this storm is officially called a hurricane.

Historically, flooding has been the most severe form of natural disaster often caused by hurricanes. The people most vulnerable to flooding live near oceans or dams on large rivers.

Meteorologists have developed a rating system for hurricanes so people will know the relative severity of a hurricane. The categories are tied to the maximum sustained wind speed in the storm.

The strongest hurricane on record was Hurricane Nancy in 1961. This storm stayed at a category 5 level for five and a half days—with winds blowing as high as 213 mph (345 kph) for over one minute at a time. Wind speeds of 90 mph can blow a car over, and wind speeds over 180 mph can lift a truck off its wheels and blow it away.[1] God's storms are powerful.

Tropical Hurricane

CHAPTER 9: GOD GIVES US ATMOSPHERE AND WEATHER

Hurricane Category	Wind Speed (MPH)	Damage at Landfall
1	74-95	Minimal
2	96-110	Moderate
3	111-129	Extensive
4	130-156	Extreme
5	157 or higher	Catastrophic

Preparing for a Major Storm

Preparedness for major storms is key, especially for those who live around the Gulf of Mexico, the United States' East Coast, the Philippines, coastal China, Japan, Mexico, Australia, Taiwan, and Vietnam. Usually, weather forecasters will give a two to three day warning before the big storm hits. The following is a good checklist for preparing for the storm.

1. Be sure that you have identified the evacuation routes out of the area.

Aftermath of Hurricane Katrina in 2005, New Orleans, Louisiana, USA

2. Keep a supply of flashlights and batteries in an easily accessible location. Consider also maintaining a good stock of non-perishable foods in storage.

3. Be aware of the flood plains. If you live anywhere near the coast, and your house may be caught in a 15-20- foot storm surge, you will need to move to higher ground before the storm hits.

4. If a storm warning has been issued, act quickly to do the following:

   a. Protect your windows with plywood or storm shutters.

   b. Secure things that could be blown around outdoors, like wheelbarrows, playground sets, outdoor patio furniture, and vehicles.

   c. Make sure you have a week's supply of food for each family member.

5. Most importantly, pray for God's protection on your family and your home.

The most dangerous thing about hurricanes is flooding. Walking or driving through fast-moving water is extremely risky. Even six inches of rushing water can sweep a man off his feet. Sometimes these flash floods can pull a car into deeper water, drowning the people inside.

Another risk during storms is carbon monoxide poisoning. When the electricity goes out, people will try to use their gas burning grills, stoves, and generators inside the house or garage. Because there isn't enough ventilation indoors to clear out the poisonous exhaust, people can die by inhaling these gases. Carbon monoxide is odorless and invisible. Burning candles and fires in an unventilated area can also be dangerous.

Tornadoes

> Behold, a whirlwind of the LORD has gone forth in fury— a violent whirlwind! It will fall violently on the head of the wicked. The anger of the LORD will not turn back until He has executed and performed the thoughts of His heart. In the latter days you will understand it perfectly. (Jeremiah 23:19-20)

Similar to hurricanes, tornadoes form when warm, humid air crashes into an area of cold air. The big difference between hurricanes and tornadoes is their size. Hurricanes can be 1,000 miles wide,

Tornado in Nebraska, USA

while the larges tornadoes are one to two miles wide. Also, tornadoes form over land, while hurricanes form over water. The winds in a tornado can move as fast as 300 miles per hour, while hurricanes tend to have winds that are less than 180 miles per hour. Tornadoes are quick to the punch and normally last less than an hour, while hurricanes can drag on for weeks.

Tornadoes typically look like a dark funnel cloud that moves quickly along the ground at 30-70 miles per hour. A tornado forms when the warm air from the ground rises up through the cold air, creating what is known as updraft—a flow of air moving from the ground up into the sky. If the updraft runs into erratic winds going every which way, this can get the draft of air to spin. This swirling action sucks more warm air up into the funnel cloud from the ground. When the draft meets cooler air, this causes the cloud to drop back to the ground, forming a tornado.

The most violent tornadoes are rated F5, and they occur about every four to seven years in the United States. The states of Alabama and Oklahoma have had the most F5 tornadoes on record (seven each), blowing at 261-318 mph. The most destructive tornado in American history took place on March 18, 1925. It swept across three states from Missouri, through southern Illinois, and into Indiana. This mile-wide, F5 twister took out 19 communities, killed almost 700 people, and traveled 235 miles in three and a half hours. About 15,000 homes were destroyed.

The most deadly tornado in the world (rated at F4), killed about 1,500 people in Bangladesh on April 25, 1989.

The United States gets about 1,000 tornadoes a year, mostly in the midwestern states. South Africa, Bangladesh, and Canada record between 50 and 100 tornadoes a year. The State of Texas takes the record for the most tornadoes in a year—up to 120 per year.

Tornadoes usually appear in the spring and summer between 3:00 pm and 9:00 pm. Some of the warning signs of an approaching tornado are:

- Dark, often greenish, sky
- Wall clouds below the storm clouds

Tornado Shelter Sign in Airport

- A cloud of debris
- Large hail without rain
- A sudden stillness before the storm
- A loud roar like a freight train signals that a tornado is only seconds away
- An obvious funnel cloud

If you notice any of these signs or hear emergency sirens or radio warnings of a tornado, act immediately.

If you are in a house, take refuge in the basement. If you don't have a basement crouch down next to an interior wall, closet, or hallway. Stay away from all windows, outside walls and doors. Hide in a bathtub, or under a heavy table. Stay away from cabinets or bookshelves that could fall on you.

Avoid sheltering in cars and mobile

homes. Rather look for shelter in a building with a solid foundation. If you can't find any other shelter, lie down in a ditch or a low-lying area far away from buildings, cars, and mobile homes. And most importantly, pray.

These powerful storms are just small examples of God's great power. Although science has given man some ability to limit the destruction, no structures can stand against an F5 tornado. Man cannot control tornadoes, nor can he predict when and where they will touch down. Only God is in control of these powerful storms. It is for us to fear God, to stand in awe of His great power. Storms serve as warnings for us; reminders that we must humble ourselves before the mighty hand of God. They remind us that God really does judge sin. These calamities are hardly comparable to the judgment He will bring upon sinners at the end of the world. Let us repent of our sins, and turn in faith to Jesus Christ, the Savior. Only He can save us from the powerful, frightening judgment on the last day. Referring to this judgment, Proverbs 1 warns of a tornado that will run down those who do not fear God.

> Wisdom calls aloud outside;
> She raises her voice in the open squares...
> "How long, you simple ones, will you love simplicity?
> For scorners delight in their scorning,
> And fools hate knowledge...
> Because you disdained all my counsel,
> And would have none of my rebuke,
> I also will laugh at your calamity;
> I will mock when your terror comes,
> When your terror comes like a storm,
> And your destruction comes like a whirlwind,
> When distress and anguish come upon you.
> Then they will call on me, but I will not answer;
> They will seek me diligently, but they will not find me.
> Because they hated knowledge
> And did not choose the fear of the LORD..."
> (Proverbs 1:20,22,25-29)

Wildfires and Forest Fires

Over the last fifty years, wildfires and forest fires have become increasingly dangerous and destructive to property. About 80% of wildfires are set by humans, usually set off by burning trash, campfires, arson, and cars or heavy equipment. Only about 20% are started by lightning strikes.

If you are ever caught in a wildfire, don't try to outrun the blaze. The safest thing to do is to hide in water—a river, a pond, or some other body of water where you can keep yourself covered mostly in the water. Look for a low lying area void of grass, bushes or trees. If possible, cover yourself with wet clothing or soil. Breathe through a moist handkerchief or cloth. Smoke inhalation is often the biggest challenge facing people caught in fires.

The better thing to do is evacuate when you hear of a wildfire in the area. If you have time before leaving your house, clear away things close to the house that might burn. That would include dry brush, firewood, barbecue propane tanks, lawnmowers containing fuel, and other combustibles. Close windows, vents, and outside doors. Turn off gas lines feeding into the house. Soak down the roof of the house with water, and fill up pools, garbage cans, and hot tubs with water—this might slow down the fire. And pray.

Fear God and Repent of Sin

When natural disasters happen, and people get killed, and property is damaged, there is a lesson to learn. When the tower fell in Siloam and killed seventeen people, Jesus responded by saying, "Unless you repent you will all likewise perish" (Luke 13:5).

The National Oceanic and Atmospheric Administration monitors the major storms, fires, flooding, and droughts that occur every year in the US. The overall damage caused by these natural disasters is getting worse with every decade that goes by. The last ten years were five times worse than what the nation experienced in the 1980s:[2]

Decade	Average Damage Cost Per Year
1980s	$20 Billion
1990s	$38 Billion
2000s	$54 Billion
2010s	$103 Billion

The money spent on rebuilding houses, businesses, roads, and bridges destroyed in these disasters continues to rise. What should Americans do? They should humble themselves before God, repent of their sins, and seek His mercy.

CHAPTER 9: GOD GIVES US ATMOSPHERE AND WEATHER

Type of Natural Disaster	Average Damage Cost Per Year
Hurricanes	$23 Billion
Flooding	$6.2 Billion
Violent Storms (Tornadoes)	$6.2 Billion
Wildfires	$2.1 Billion
Winter Storms	$1.2 Billion

Weather Forecasting

Ultimately, God is in control of the weather. Small changes in sun activity can bring about warming of certain parts of land and ocean. It's impossible to know for sure what the weather will be one week from now. However, meteorologists are somewhat adept at forecasting the weather up to three days ahead of time. These scientists can come up with a good guess about what will occur in a certain area at a certain time, but they can't be certain about what will happen in the future.

Meteorologists use special balloons in different locations to identify pressures and temperatures up in the atmosphere. These balloons can rise to 24 miles high (39 km), well into the stratosphere.

Weather satellites help meteorologists watch cloud movements around the world. They don't always know where the clouds will move next, because that depends on the winds, and only God is in control of the wind. No computer modeling can figure out for sure where the winds are going to blow tomorrow or the next day.

Barometers are helpful to predict weather in the near future, over the next 2–12 hours. If the atmospheric pressure is relatively high, weather will probably be sunny and stable. If the pressure drops, you can expect rainy, or even stormy weather.

Thousands of years ago, people would try to predict the weather by the clouds. But in Matthew 16, Jesus was more interested in a spiritual sensitivity to the

SUN	MON	TUE	WED	THU	FRI	SAT
68	74	83	75	82	81	90
WINDY	SUNNY	THUNDERSTORMS IN THE AFTERNOON	MOSTLY CLOUDY	PARTLY CLOUDY	RAIN	SUNNY

Weather Forecast

GOD MADE THE WORLD

Weather Balloon

sins of a nation, and the judgment of God due to us for sin.

Then the Pharisees and Sadducees came, and testing Him asked that He would show them a sign from heaven. He answered and said to them, "When it is evening you say, 'It will be fair weather, for the sky is red'; and in the morning, 'It will be foul weather today, for the sky is red and threatening.' Hypocrites! You know how to discern the face of the sky, but you cannot discern the signs of the times." (Matthew 16:1-3)

What Have We Learned?

Consider everything you have learned about this created world throughout this study. Surely, this is a spectacular, complex, beautiful, astonishing, extraordinary, incomprehensible, wondrous, and thrilling creation. But please do remember that the Lord Jesus Christ is the Creator of all of the awe-inspiring universe and our beautiful earth. But, He also sustains all things by the Word of His power. He keeps it going. He rules over all of nature. Nothing happens in the weather, in the fault lines under the crust of the earth, with asteroids or comets, or anything else, without His permission. We see His mercy as He sustains the world with water, air, and material resources. We also witness God's judgment upon the earth when natural disasters happen, and when disease and death occur.

Hopefully, young readers, you have

CHAPTER 9: GOD GIVES US ATMOSPHERE AND WEATHER

learned more humility as you have studied God's awesome world. By this time, you have come across many more reasons why the fear of God is the beginning of knowledge. God is good. He is powerful. His wisdom is above and beyond what we could ever understand. Now, you have more reasons to praise Him for His wonderful works all around you! ▪

The twenty-four elders fall down before Him who sits on the throne and worship Him who lives forever and ever, and cast their crowns before the throne, saying:
"You are worthy, O Lord,
To receive glory and honor and power;
For You created all things,
And by Your will they exist and were created." (Revelation 4:10-11)

Isle of Skye, Scotland, UK

Pray
- Thank the Lord for the atmosphere, for clouds, and especially for rain that waters the earth.
- Praise Him for His mighty works that we see all around us. Praise Him for the mighty winds, the hail and the snow, the storms and the calm.

Sing
Hallelujah, Praise Jehovah, From the Heavens Praise His Name (Psalm 148)

Hallelujah, praise Jehovah,
From the heavens praise His name;
Praise Jehovah in the highest,
All His angels praise proclaim.
All His hosts together praise Him,
Sun, and moon, and stars on high;
Praise Him, O ye heav'n of heavens,
And ye floods above the sky.

Refrain:
Let them praises give Jehovah,
For His Name alone is high,
And His glory is exalted,
And His glory is exalted,
And His glory is exalted,
Far above the earth and sky.

Let them praises give Jehovah,
They were made at His command,
Them forever He established;
His decree shall ever stand.
From the earth, O praise Jehovah,
All ye floods, ye dragons all;
Fire, and hail, and snow, and vapors,
Stormy winds that hear His call.

All ye fruitful trees and cedars,
All ye hills and mountains high,
Creeping things, and beasts, and cattle,
Birds that in the heavens fly.
Kings of earth, and all ye people,
Princes great, earth's judges all,
Praise His name, young men and maidens,
Aged men and children small.

If you do not know this psalm, you may listen to a version of the hymn on the Internet, with supervision, and sing along with it.

Do
1. **Build your own weather station**, equipping yourself with a rain gauge, a barometer, a wind vane, a hygrometer (for humidity measurements), and an anemometer

(wind gage). Check out youngexplorers.com for supplies. Track changes in temperature, pressure, humidity, wind, and precipitation over seven days, logging your measurements twice a day. Try to predict the weather for the next day based on your measurements. Consider joining the weatherunderground team of 180,000 personal weather stations.

2. **Make a winter survival kit for your car.** This activity would be especially helpful for those living in colder climates. This would include, at a minimum, a flashlight, a blanket, a small shovel, batteries, snacks, water, gloves, boots, and a first-aid kit.

3. **Make sure your home is prepared for severe weather.** Prepare for electrical outages, snow storms, tornadoes, or hurricanes by going over the following checklist:

- ❏ Adequate flashlights and batteries in some place where you can find them
- ❏ An alternate place to cook (preferably outdoors or where there is good ventilation)
- ❏ Fuel for cooking or heating. This could include wood, kerosene, Sterno, or other types of fuel that are safe to store.
- ❏ A supply of non-perishable foods
- ❏ Matches or lighters in some place where you can find them
- ❏ First aid kit and basic medications where you can find them
- ❏ A source of fresh water or a supply of bottled water
- ❏ A bucket to carry water needed for flushing toilets
- ❏ A transistor radio with batteries
- ❏ A safe way to escape flooding or heavy tornado winds for those living in at-risk areas
- ❏ Coolers for food storage in case the electricity goes out. You could use ice from the freezer to keep the food cooler, longer.

Watch

To watch the recommended videos for this chapter, go to **generations.org/ GodMadeTheWorld** and scroll down until you find the video links for Chapter 9. Our editors have been careful to avoid films with references to evolution. However, we would still encourage parents or teachers to provide oversight for all internet usage. These videos may not give God the glory for His amazing creative work, so the student and parent/teacher should respond to these insights with prayer and praise.

Victoria Falls in Africa

ENDNOTES

Chapter 1
1. Gilbert King, "How Edwin Hubble Became the 20th Century's Greatest Astronomer," Smithsonian Magazine, May 20, 2013, https://www.smithsonianmag.com/history/how-edwin-hubble-became-the-20th-centurys-greatestastronomer-66148381/.
2. "Scientists and Belief," *Pew Research Center*, November 5, 2009, https://www.pewforum.org/2009/11/05/scientists-and-belief/.

Chapter 2
1. Quoted in David Brewster, *Memoirs of the Life, Writings, and Discoveries of Sir Isaac Newton* (Edinburgh: Thomas Constable and Co., 1855), 2:407.

Chapter 3
1. "Why We Explore," NASA, https://www.nasa.gov/exploration/whyweexplore/why_we_explore_main.html.
2. Elaina Zachos, "Antarctica Was Once Covered in Forests. We Just Found One That Fossilized," *National Geographic*, November 15, 2017, https://www.nationalgeographic.com/news/2017/11/ancient-fossil-forest-found-antarctica-gondwana-spd/.
3. Matt Walker, "Toads can 'predict earthquakes' and seismic activity," BBC Earth News, March 31, 2010, http://news.bbc.co.uk/earth/hi/earth_news/newsid_8593000/8593396.stm.
4. "Global Energy Sources," *Pennsylvania State University*, https://www.e-education.psu.edu/earth104/node/1345.
5. "Countries with the biggest coal reserves," *Mining Technology*, January 6, 2020, https://www.mining-technology.com/features/feature-the-worlds-biggest-coal-reserves-by-country/.
6. Nick Cunningham, "U.S. Has World's Largest Oil Reserves," *OilPrice.com*, July 5, 2016, https://oilprice.com/Energy/Energy-General/US-Has-Worlds-Largest-Oil-Reserves.html.
7. "Electric Generating Costs: A Primer," *Institute for Energy Research*, August 22, 2012, https://www.instituteforenergyresearch.org/renewable/electric-generating-costs-a-primer/.
8. "Renewable Power Generation Costs in 2018," *International Renewable Energy Agency*, May 2019, https://www.irena.org/publications/2019/May/Renewable-power-generation-costs-in-2018.

Chapter 4
1. Robert Boyle, *The Works of the Honourable Robert Boyle* (London: J. and F. Rivington, 1772), 1:262.
2. Ibid., 5:558.

Chapter 5
1. Isaac Newton, *Newton's Principia*, trans. Andrew Motte (New York: Daniel Adee, 1848), 504.
2. William H. Thomson, *Christ in the Old Testament, or The Great Argument* (New York: Harper, 1888), 459.
3. Quoted in Lewis Campbell and William Garnett, *The Life of James Clerk Maxwell* (London: Macmillan, 1882), 323.
4. Benjamin Franklin, *The Works of Benjamin Franklin* (Philadelphia: William Duane, 1809), 6:241.

Chapter 7
1. Quoted in Geoffrey Cantor, *Michael Faraday: Sandemanian and Scientist* (London: Macmillan, 1991), 200.
2. Ibid., 81.
3. Quoted in Walter Baxendale, *Dictionary of Anecdotal Incident, Illustrative Fact* (New York: Thomas Whitaker, 1888), 290.

Chapter 8
1. Lester R. Brown, "Full Planet, Empty Plates: The New Geopolitics of Food Scarcity," *Earth Policy Institute*, http://www.earth-policy.org/books/fpep/fpepch7.
2. Quoted in John Perry, *Unshakable Faith: Booker T. Washington & George Washington Carver* (Sisters, OR: Multnomah, 1999), 94.

Chapter 9
1. Thomas W. Schmidlin, et al., "Wind Speeds Required to Upset Vehicles," *The Conference Exchange*, https://ams.confex.com/ams/pdfpapers/50675.pdf.
2. "Billion-Dollar Weather and Climate Disasters: Time Series," *National Oceanic and Atmospheric Administration*, https://www.ncdc.noaa.gov/billions/time-series [numbers adjusted for inflation].

ENDNOTES

Aurora Borealis (Northern Lights) Viewed in Alaska

LIST OF IMAGES

Introduction
1. Haifoss Waterfall, Iceland | iStock.com
2. Great Ocean Road, Victoria, Australia | iStock.com

Chapter 1
1. Tian Shan Mountain Range, Asia | iStock.com
2. Logs in a Pine Forest | iStock.com
3. View of Earth from the Moon | iStock.com
4. Andromeda Galaxy | iStock.com
5. Milky Way Galaxy | iStock.com
6. Sparrow | iStock.com
7. Noah's Ark and the Worldwide Flood | iStock.com
8. Monarch Butterfly | iStock.com
9. Grapevine | iStock.com
10. Child's Nature Notebook | iStock.com
11. Child Using Magnifying Glass | iStock.com

Chapter 2
1. Saturn | iStock.com
2. Buzz Aldrin on the Moon, 1969 | Wikimedia Commons | Public Domain
3. Milky Way Galaxy | iStock.com
4. Planets in the Solar System | iStock.com
5. Mars | iStock.com
6. The Sun's Nuclear Fusion Reaction | iStock.com
7. Stars in the Milky Way Galaxy | iStock.com
8. Sunlight Lightens Planets | iStock.com
9. Simulated View of a Black Hole | iStock.com
10. Nebula | iStock.com
11. Great Nebula Surrounding Eta Carinae | iStock.com
12. Artistic Rendering of a Quasar | iStock.com
13. Great White Spot on Saturn | iStock.com
14. Clouds of Jupiter | iStock.com
15. Sun Illuminates Earth and Moon | iStock.com
16. Scorpius the Scorpion Constellation | iStock.com
17. Leo the Lion Constellation | iStock.com
18. Earth Tilts to the Sun | iStock.com
19. Sydney, Australia in the Springtime | iStock.com
20. The Great Wall of China in the Fall | iStock.com
21. Conditions near the Equator in the Democratic Republic of the Congo | iStock.com
22. Phases of the Moon | iStock.com
23. Solar Eclipse | iStock.com
24. Lunar and Solar Eclipse | iStock.com
25. Spilled Coffee | iStock.com
26. Planets in Orbit | iStock.com
27. Astronauts Floating in Space | iStock.com
28. Walking on the Moon (Elements Furnished by NASA) | iStock.com
29. Milky Way | iStock.com

GOD MADE THE WORLD

Chapter 3
1. The Earth | iStock.com
2. Antarctica | iStock.com
3. Noah's Ark | iStock.com
4. Child Digging a Hole | iStock.com
5. Structure of the Earth | iStock.com
6. Ring of Fire | Wikimedia Commons | Public Domain
7. Mount Bromo, East Java, Indonesia | iStock.com
8. Seismograph | iStock.com
9. 2015 Illapel Earthquake Aftermath | iStock.com
10. Tectonic Plates | iStock.com
11. Earthquake Epicenter | iStock.com
12. Aftermath of the 2004 Tsunami in Indonesia | Wikimedia Commons | Public Domain
13. Tungurahua Volcano Erupts in Ecuador | iStock.com
14. Mount Rainier, Washington | iStock.com
15. Mount Vesuvius and the Ruins of Pompeii | iStock.com
16. Highway After Earthquake | iStock.com
17. Foundations of a House | iStock.com
18. 1980 Eruption of Mt. St. Helens | Wikimedia Commons | Public Domain
19. Volcanic Lava | iStock.com
20. Himalayan Mountains, Nepal | iStock.com
21. Paria River Valley, Utah | iStock.com
22. Mount Everest, Nepal | iStock.com
23. Igneous Rock | iStock.com
24. Sedimentary Rock | iStock.com
25. Grand Canyon, Arizona | iStock.com
26. Comet | iStock.com
27. Asteroid | iStock.com
28. Meteorite | iStock.com
29. Meteor Crater, Arizona, USA | iStock.com
30. Dinosaur Fossil in Rock | iStock.com
31. Oil Pump Jack in Alberta, Canada | iStock.com
32. Trans-Alaska Pipeline | iStock.com
33. Tihange Nuclear Power Station in Huy, Belgium | iStock.com
34. Green Pine Forest | iStock.com

Chapter 4
1. Yosemite Valley, California | iStock.com
2. Bowling Ball | iStock.com
3. Ping Pong Ball | iStock.com
4. Diagram of Atom | iStock.com
5. Molecule, Atom, and Parts of the Atom | iStock.com
6. Without Gravity, Everything Floats Away | iStock.com
7. Sand | iStock.com
8. Legos | iStock.com
9. Stainless Steel Pan | iStock.com
10. Tree | iStock.com
11. Dog | iStock.com
12. Iodine Vapors | iStock.com
13. Periodic Table of Elements | iStock.com
14. Pan with Teflon Coating | iStock.com
15. Baking Soda | iStock.com
16. Pan with Water Turning into Steam | iStock.com
17. Robert Boyle (1627-1691) | Wikimedia Commons | Public Domain
18. Smoothie Mixture | iStock.com
19. Rusted Truck | iStock.com
20. Salt | iStock.com
21. Sugar in Iced Tea | iStock.com
22. Soda Drinks | iStock.com
23. Orange Juice | iStock.com
24. Coffee Drink Mixture | iStock.com
25. The Wedding at Cana | iStock.com
26. Covalent Bond | iStock.com
27. Diagram of Water | iStock.com
28. Pillow | iStock.com
29. Cement Brick | iStock.com
30. Hardwood Floor | iStock.com
31. Boy on Scale | iStock.com

298

LIST OF IMAGES

32. Moon Surface | iStock.com
33. Cutting Bricks | iStock.com
34. Paper | iStock.com
35. Drilling Through Sheet Metal | iStock.com
36. Rope | iStock.com
37. High Tensile Strength Steel Wire | iStock.com
38. Boeing 747 Jet | iStock.com
39. Charles Martin Hall | Wikimedia Commons | Public Domain
40. Cement Mixture in Wheelbarrow | iStock.com
41. Excavator at a Mine | iStock.com

Chapter 5
1. Wind Turbines | iStock.com
2. Moving Heavy Objects is Work | iStock.com
3. Slingshot Pulled Back | iStock.com
4. Boy Reading with Flashlight | iStock.com
5. Sir Isaac Newton | Wikimedia Commons | Public Domain
6. Coasting on Bikes | iStock.com
7. Filling Tank with Gas | iStock.com
8. Heat Energy Transfers to Motion | iStock.com
9. Lord Kelvin | Wikimedia Commons | Public Domain
10. Gaps in Sidewalk | iStock.com
11. Wooden Spoon in Pot | iStock.com
12. Radiative Heater | iStock.com
13. Toaster with Bread | iStock.com
14. Statue of James Clerk Maxwell | Wikimedia Commons | Public Domain
15. Magnet and Steel Balls | iStock.com
16. The Magnetic Field | iStock.com
17. Earth's Magnetic Field | iStock.com
18. Lightning Strike in El Paso, Texas | iStock.com
19. Benjamin Franklin | Wikimedia Commons | Public Domain
20. Light Bulb | iStock.com

21. Balloon Sticks to Wall | iStock.com
22. Electric Pylons | iStock.com
23. Child Safety Cover on Outlets | iStock.com
24. Electrical Shorts Cause Fires | iStock.com
25. Circuit Breaker Box | iStock.com
26. Electricity Insulators | iStock.com
27. Electrical Bill | iStock.com
28. Digital Thermostat | iStock.com
29. Packed Freezer | iStock.com
30. Water Heater Temperature Control | iStock.com
31. Stars in Sky | iStock.com

Chapter 6
1. Car Crossing the Bixby Canyon Bridge, Big Sur, California | iStock.com
2. Wagon Ride | iStock.com
3. Bicycle Ride | iStock.com
4. Mountain Biking Down a Mountain | iStock.com
5. Demonstration of Force Required to Push a Car Uphill | iStock.com
6. Bicycling Through Mud and Sand | iStock.com
7. Bugatti Veyron | iStock.com
8. Gravity Causes Leaves to Fall to the Ground | iStock.com
9. Basketball Hoop | iStock.com
10. Brakes Slow Down Bikes | iStock.com
11. Front Wheel of a Bicycle | iStock.com
12. Car Engine | iStock.com
13. Combustion Engine Diagram | iStock.com
14. Newton's Pendulum | iStock.com
15. Orville Wright and Wilbur Wright | Wikimedia Commons | Public Domain
16. First Flight at Kitty Hawk | Wikimedia Commons | Public Domain
17. Jet Airplane | iStock.com
18. Diagram of Wing | iStock.com
19. Duck | iStock.com

20. Bernoulli Principle | iStock.com
21. Space Shuttle | iStock.com
22. Balloon Rocket | Wikimedia Commons | Public Domain
23. Plane Taking Off Runway

Chapters 7
1. Tulip Fields | iStock.com
2. Street Lights | iStock.com
3. Mountain Biking with Head Lamp | iStock.com
4. Oldest Surviving Photograph | Wikimedia Commons | Public Domain
5. Ripples in Water | iStock.com
6. Glass Cup | iStock.com
7. Light Spectrum | iStock.com
8. Rainbow Over Manhattan, New York City | iStock.com
9. Colored Pencils | iStock.com
10. Prism Demonstration | iStock.com
11. Light Beams Traveling Through Mist | iStock.com
12. Lady Reflected in Mirror | iStock.com
13. Diagram of Mirror Reflection | Wikimedia Commons | Public Domain
14. Albert Einstein | iStock.com
15. Solar Panels | iStock.com
16. Sound Wave | iStock.com
17. Michael Faraday | iStock.com
18. Loudspeaker Demonstration | iStock.com
19. Frequency of Sound | iStock.com
20. Microphone | iStock.com
21. Dolphins Use Echo-Location to Gage Distances | iStock.com
22. Ultrasound of a Baby in Mother's Womb | iStock.com
23. Concert Hall | iStock.com
24. Child Playing Violin | iStock.com
25.

Chapter 8
1. Lake Tahoe, California | iStock.com
2. Greenland | iStock.com
3. Buoyant Force | iStock.com
4. Kayaking | iStock.com
5. Container Ship | iStock.com
6. Sea of Galilee | iStock.com
7. Firefighter with Fire Hose | iStock.com
8. Scuba Diving | iStock.com
9. Watering the Garden | iStock.com
10. Photosynthesis | iStock.com
11. Snow Near Crested Butte | iStock.com
12. Snowflake Photos Taken by Wilson Bentley | Wikimedia Commons | Public Domain
13. Children on a Toboggan | iStock.com
14. Child Playing in the Dirt | iStock.com
15. Transplanting a Plant | iStock.com
16. Seedlings in Fertile Aoil | iStock.com
17. Sandy Soil | iStock.com
18. Soil Layers | iStock.com
19. Tractor Plowing Field | iStock.com
20. Soybean Field | iStock.com
21. George Washington Carver | Wikimedia Commons | Public Domain
22. Rice Fields in Vietnam | iStock.com

Chapter 9
1. Storms in Ireland | iStock.com
2. Atmosphere Layers | iStock.com
3. Clouds in the Troposphere | iStock.com
4. A Bottle at Various Elevations | Image by Author
5. Car in Mountain Pass | iStock.com
6. Barometer | iStock.com
7. Tornado | iStock.com
8. Lighting | iStock.com
9. Heavy Wind During a Hurricane | iStock.com

10. The Water Cycle | iStock.com
11. Rain in Kerala, India | iStock.com
12. Nimbostratus Clouds | iStock.com
13. Large Hailstones | iStock.com
14. Hailstorm in Tennessee | iStock.com
15. Dew on Grass | iStock.com
16. Lightning Above Ft. Worth | iStock.com
17. Sahara Desert, Morocco | iStock.com
18. Snow-Covered Highway in Austria | iStock.com
19. Snow Igloo | iStock.com
20. Tropical Hurricane | iStock.com
21. Aftermath of Hurricane Katrina | iStock.com
22. Tornado in Nebraska | iStock.com
23. Tornado Shelter Sign in Airport | iStock.com
24. Weather Forecast | iStock.com
25. Weather Balloon | iStock.com
26. Isle of Skye, Scotland | iStock.com

Endnotes

Victoria Falls in Africa | iStock.com

List of Images

Aurora Borealis (Northern Lights) Viewed in Alaska | iStock.com

Index

Fiji Islands | iStock.com

Fiji Islands

INDEX

A

Absorb, 155, 202, 208–210, 213, 223, 239, 242, 261
Acceleration, 55, 181–183, 185–186
Acoustics, 223
Actinomycetes, 245
Action, 143, 177, 190–191, 193, 196, 238, 239
    Capillary Action, 239
Adam, 64, 224, 249
Agur, 20
Air, 12, 15, 19, 36, 52, 57, 64, 87, 93, 114, 116, 146, 152–154, 156, 170, 177, 179, 183, 186, 189, 191–196, 206–209, 212, 216–219, 229, 237, 239, 242, 247, 248, 260, 262, 264–267, 269–271, 273–274, 276, 283, 284, 289
    Air Masses, 267
    Air Resistance, 146, 183, 196
Airplane, 29, 133, 192–197, 203, 221, 260, 279
Alkali Metals, 108
    Alkali Earth Metals, 108
Alkanes, 93
Altitude, 262–264
Aluminum, 133–134, 136, 153, 168, 215
Amplitude. *See Waves*
Animals, 14, 17, 64–65, 91–92, 106, 156, 160–161, 178, 219, 243, 269
Apollo 11, 26, 196
Argon, 108, 116
Ash, 80–81
Astatine, 108, 112
Asteroids, 32, 88–89, 289
Astronomers, 15–16, 34, 40, 89, 90
Atmosphere, 30, 32, 35, 36, 69, 88–89, 154, 206–207, 260–264, 270, 287
Atom, 33, 77, 102–104, 106–108, 112–113, 116–117, 123, 126, 144, 148–149, 152, 154, 157, 162–163, 168, 201, 214, 231, 234, 260
Automobile, 30, 57, 78–79, 81, 93–95, 98, 102, 105, 107, 117–118, 127, 131–132, 135–136, 143–144, 146–148, 156, 173, 178–179, 181–185, 187–190, 193, 196–197, 203, 263–264, 278–281, 282, 285–286
Axle, 188–189

B

Bacteria, 245
Ball Bearings, 188
Barometer, 265, 287
Battery, 128, 144, 161–162, 165, 282
Bauxite, 133
Bends, 237

Bentley, Wilson, 241
Bernoulli Principle, 98
Bicycle, 131, 178–182, 187, 189, 264
Birds, 14, 153, 160, 185, 192, 223, 269
Black Hole, 37, 40
Boat, 178, 190–193, 231–233, 259
Boyle, Robert, 115, 136–137, 145
    Boyle's Law, 115
Bricks, 78, 102, 107, 126, 129, 132, 149–150, 153
Bromine, 108
Buoyancy, 231–233
Butane, 93
Byproduct, 156

C

Calcium, 108, 136, 245, 249
Camera, 204, 241
Canis Majoris, 33–34
Canyons, 85–86
Capacitors, 165
Capillary Action. *See Action*
Carbon, 91–93, 109, 116, 119–121, 131–132, 239, 249, 279, 282
    Carbon Dioxide, 116, 119, 121, 239
Carver, George Washington, 253
Casting, 91, 153
Catastrophe, 68, 88
Cement, 134, 136
Ceres, 31
Chalk, 129, 247, 249–250
Chemical Analysis, 115
Chemical Bond, 104, 106
Chemical Equation, 124
Chemical Process, 117, 123, 134
Chemistry, 115, 136, 145, 219
Chlorine, 108, 245

Circuit, 165–167, 216
    Short Circuit, 168
Clouds, 14, 38, 41, 81, 124, 161, 241, 242, 260, 262, 267, 269–271, 274, 280, 283–284, 287, 288
    Cloud Type, 270
    High-level Clouds, 271
    Low-level Clouds, 271
    Mid-level Clouds, 271
Coal, 92–96, 98, 133, 153–154, 160–161, 168
    Metallurgical Coal, 133
Cobalt, 89, 158
Color, 21, 38, 107, 108, 204, 207–211, 214, 243
Comet, 87–88, 289
Compound, 108–109, 112, 115–117, 123, 133
Concentration, 118, 120
Concrete, 80, 134–135, 151, 153, 213
Condensation, 269
Conduction, 152–154
    Conductors, 153, 168, 215
    Semi-Conductor, 215
Constellations, 40, 44, 47
Contraction, 151, 231, 237
Convection, 152, 154
Copper, 93, 133, 153, 163, 168, 245
Covalent Bond, 123–124
Creation, 13, 15
Current, 133, 144, 162–167, 221, 242
Cylinder, 148, 151, 189

D

Darkness, 40, 43, 46, 49, 152, 201
Death, 59, 64, 70, 82, 165, 217, 224, 249, 289
Decibels, 221–222
Dehydration, 276
Democritus, 15
Density, 126–128, 232–233

INDEX

Desert, 64–65, 215, 275
Dew, 272–273
Diamond, 129, 132, 135, 153
Dirt, 12, 20, 52, 66, 87, 106, 146, 155, 229, 243, 252
Disaster, 69, 73, 75, 77, 79, 81, 90, 197, 280, 286–287, 289
Displacement, 232–233
Doping, 215

E

Earth, 12, 14, 17, 19, 25–35, 37, 39–45, 48–49, 51–57, 63–69, 72–73, 75–76, 78, 82–85, 87–90, 91–94, 96, 103–104, 106, 122, 125–127, 131–132, 136, 146, 154–156, 158–160, 168, 173, 185–186, 196, 202, 206–207, 229–230, 236–238, 242–243, 247, 249–250, 253, 260–262, 266–269, 271, 273, 274, 288–289
    Core, 66–67, 158–159
    Crust, 67–68, 72, 76, 83, 87, 88, 93, 96, 288
    Mantle, 67, 84, 88, 96
Earthquakes, 17, 68–74, 76–80
Echo, 223
    Echo Location, 223
Eclipse, 48–49, 50
Edison, Thomas, 162
Einstein, Albert, 147, 155, 214
Elbe Mountain Range, 84
Electricity, 81, 95–96, 161–166, 168–169, 172, 178, 189, 214–215, 276, 282
    Current Electricity, 163
    Static Electricity, 162–163, 274
Electromagnetism, 51, 156
    Electromagnetic Disturbance, 202
    Electromagnetic Scale, 205
Elements, 89, 107–108, 110, 112, 115, 123, 126, 133, 158, 277
Elliptical, 44
Energy, 27, 33, 36, 40, 59, 75, 92–97, 114, 123, 141–149, 152, 154–157, 160–165, 168–169, 171–173, 178–180, 182, 184, 201–202, 204–205, 209–210, 214–216, 221–222, 238–240, 261, 268, 274, 276
    Electrical Energy, 144, 157, 160–162, 165, 216, 221–222
    Electromagnetic Energy, 155, 205
    Geothermal Energy, 95–96, 144
    Kinetic Energy, 143, 145–146, 149, 162
    Nuclear Energy, 96
    Potential Energy, 143
    Thermal Energy, 149
    Types of Energy, 144
Engine, 95, 144, 147–148, 151, 154, 173, 181–182, 184, 189, –191, 193, 221
Epicenter, 72
Equator, 30, 44–47, 65–66, 280
Equinox, 46–47
Eris, 31
Ethane, 93
Evaporation, 269
Eve, 64
Evil, 82, 224
Expansion, 151
Eyes, 13, 16, 21, 87, 210–212, 214, 216, 218, 262

F

Faith, 83, 259, 285
Faraday, Michael, 215, 217
Fear of God, 22, 58, 69, 78, 90, 197, 204, 259, 274, 285–286, 289
Fiber, 135, 153, 204, 242
    Fiberglass, 153, 242
Fingernail, 129
Fire, 33, 35, 78, 81, 98, 123, 155–156, 166, 202, 236, 265–266, 274, 276, 282, 286–287
    Firewalking, 153

Wildfire, 265, 286–287
Fluorine, 108–109
Force
    Centripetal Force, 53–54
    Electromagnetic Force, 103–104, 106, 156–157, 161, 201
    Friction Force, 180
    Gravitational Force, 37, 51, 55–56, 106, 184
    Nuclear Force, 51, 105
    Strong Force, 104–106, 157
    Weak Force, 104, 157
Fossil, 65, 91–92
Franklin, Benjamin, 161–162
Frequency. *See Waves*
Friction, 146, 179–183, 186–188
Front, 267
    Cold Front, 267
    Warm Front, 267
Frost, 272–273
Fuel, 33, 92–95, 128, 148, 168, 173, 184, 189–190, 193, 196, 286
Fungi, 245

G

Galaxy, 15–16, 27–29, 184
    Milky Way, 15–16, 28–29, 36, 40
Gamma Waves, 206
Gas, 25, 27, 33, 38, 76–77, 87, 93–96, 108, 112–116, 119, 123, 124, 132, 147–148, 152, 154, 156, 170, 178, 189–191, 193, 196, 217, 231, 236–237, 241, 243, 260–262, 264, 269, 273, 278, 282, 286
    Noble Gas, 108
Gasoline, 95, 147–148, 178, 189–190, 196
Glass, 20–21, 52, 119–120, 129, 133, 135, 153, 168, 207, 208, 213–214, 223, 269
Glucose, 240

Gold, 93, 112, 128–129, 168
Government, 63, 81, 197
Graphene, 131, 135
Gravity, 28, 32, 51–57, 104–106, 126–127, 157, 180, 185–186, 194, 201, 236–237, 239, 261–262

H

Hail, 267, 271–272, 284
Hair, 103, 131, 135, 167, 214
Hall, Charles Martin, 134, 136
Halogens, 108
Hardness, 127, 129
Haumea, 31
Heat, 29, 45, 67, 84, 91, 93–96, 113–114, 118–119, 123, 132–133, 142, 144, 147–156, 160–162, 165, 168–171, 173, 189, 202, 209, 210, 231, 248, 261, 265–267, 274–275, 277–279
    Heat Transfer, 151–152
Heaven, 14, 18, 25–26, 34, 40, 55, 64, 67, 217, 238
Helium, 33, 108, 114, 116
Hemispheres, 43–46
Hertz, 219
Hexane, 93
Himalaya Range, 85
Horsepower, 183–184
Human Bones, 102, 135, 186
Humidity, 272–273
Humility, 21–22, 27, 78, 145, 214, 241, 253, 285, 287–288
Hurricanes, 265, 279–283, 287
Hydrocarbons, 93
Hydrogen, 33, 93, 108, 112, 116, 123–124, 234
Hyper-novas, 36
Hypergiant Stars, 38–40
Hypothermia, 277
Hypothesis, 145

INDEX

I

Ice, 87, 113, 119–120, 187, 231–232, 270–271, 277
Implosion, 37
Infrared, 206
Insects, 14, 245
Insulators, 153–154, 165, 168, 170–171, 215, 231, 242, 279
Iodine, 108, 116
Iron, 25, 66, 89, 93, 114, 117–118, 126–127, 132–133, 136, 153, 158, 219, 232–233, 245
    Iron Ore, 132–133

J

Jesus, 49–52, 58, 67, 69, 71, 78–79, 83, 89, 94, 104–106, 121, 150, 204, 212–213, 224, 230, 234–235, 253, 259, 285–286, 288
Joshua, 50, 273
Judgment, 67, 69, 76–78, 90, 161, 285, 288–289
Jungle, 64–65
Jupiter, 29–30, 32, 41, 55–57, 173, 186

K

Krypton, 116

L

Lake, 12, 64, 80, 82, 118, 124, 158, 229–232
Lava, 81
Law, 65, 115, 125, 144–145, 182, 185, 190–191, 197
Lead, 114, 126, 128
Leverage, 179
Lift, 193–195
Light, 27, 29, 35, 36–43, 45, 49, 55, 59, 64, 114–115, 144, 147–148, 152, 154–156, 165–166, 201–216, 218–219, 223, 243, 270, 276
Light-years, 15, 27, 36, 38
Lightning, 144, 158, 161–162, 265, 274–275, 286
Limestone, 84, 134
Liquid, 93, 112–114, 118–119, 125, 132, 147, 151–152, 154, 196, 222, 237, 241, 262, 269, 273, 277
Load, 165, 166
Lodestones, 158
Lord Kelvin, 149–150
    Thomson, Williams, 149
Lubrication, 188

M

M87, 184
Magma, 72, 76–77, 84, 87
Magnesium, 108, 245
Magnetism, 156–158
Makemake, 31
Mars, 25, 29, 31–32, 55–56, 63, 87, 186
Mass, 25, 29, 35, 37, 39, 53, 55, 69, 83, 102, 113–114, 126–127, 146–149, 173, 183, 201–202, 262, 267, 270
Matter, 93, 102, 106, 109, 112–113, 125, 141, 147–148, 202
Maxwell, James Clerk, 155, 202
Mechanical Advantage, 178–179
Medicine, 20, 253
Mercury, 25, 29, 31–32, 56, 128, 151, 184
Mercy of God, 136, 154, 287, 289
Metal, 25, 88–89, 98, 108, 128–129, 132–133, 166, 233
Meteorites, 88–90
Meteoroids, 88–89
Meteorologists, 265, 271, 280, 287
Methane, 93, 116
Microorganisms, 243, 245
Microphone, 221–222
Microwave, 144, 155, 167, 206
Minerals, 103, 243–245, 247
Mirror, 212–214
Mixture, 115, 116, 118–120, 121, 133, 134, 136, 240, 249
Moh's Hardness, 129

307

Moisture, 14, 156, 251, 252, 270–273
Molecule, 103, 106–107, 109, 112–113, 116, 119, 123–124, 149, 232, 239, 260–262, 265, 270
Moon, 12, 14, 17, 25, 26, 28, 34, 41–43, 48–51, 53–57, 64, 84, 87–88, 90, 106, 127, 185–186, 196, 203–204, 262, 270
    Full, 48
Motion, 113, 143, 145–147, 162, 182, 190–191, 194, 196, 205
Motors, 132, 165, 178–179, 189
Mountain, 12, 17, 58, 64, 74, 76, 80–85, 89, 92, 142, 203, 212, 262–263
Music, 218, 222, 225

N

Nebula, 37–38
Nematodes, 245
Neodymium, 158
Neon, 108, 116
Neptune, 29, 31, 56
Newton, Isaac, 57, 145–147, 182–183, 190, 207
    Newton's Law of Motion
        First, 145–146
        Second, 182
        Third, 190
Newton's Pound-Force, 55
Nickel, 89, 158, 245
Niépce, Joseph Nicéphore, 204
Nitrogen, 116, 242, 245
    Nitrogen Dioxide, 116
Noah, 51, 69, 190, 232
Noble Gas. *See Gas*
North Pole, 43, 66, 156, 158–159
Northern and Southern Lights, 261
Nuclear
    Nuclear Bomb, 33, 75, 80, 83, 90, 150
    Nuclear Fusion, 33, 96, 148
    Nuclear Reactors, 75
Nucleus, 103–106, 113, 126, 157

O

Ocean, 12, 15, 17, 56, 58, 64, 68, 70, 74, 85–87, 89, 92, 205–206, 229–231, 269, 280, 286–287
Oil, 12, 92–94, 98, 118, 130, 147, 168, 187, 247
Organic Materials, 243–245, 247–249, 251
Osmium, 126, 128
Oxygen, 103, 107, 108, 112, 114, 116–118, 123–124, 132–133, 193, 196, 240, 260
Ozone, 112, 116, 260–261

P

Palladium, 128
Pentane, 93
Periodic Table, 108, 110
Petrifaction, 91
Phases, 48–49
Photo-electric Cells, 214
Photography, 128, 204, 241
Photons, 201–202, 205, 213–214, 216
Photosynthesis, 239
Physics, 141, 145, 181
Pitch, 219
Planet X, 30
Plants, 17, 19–20, 64, 95, 106, 154, 231, 239–240, 242–245, 247–253, 269
Plate Tectonics, 33
Plunkett, Roy J., 109, 136
Pluto, 29, 31, 55–56
Plutonium, 128
Potassium, 108, 245
Pressure, 66–67, 76, 91, 93, 115, 148, 164, 180, 189–190,

INDEX

193–195, 236–237, 262–267, 269, 276, 287, 288
    Air Pressure, 194, 262, 264–266, 274
    Atmospheric Pressure, 262–265, 287
    Underwater Pressure, 237
    Water Pressure, 236–237
Pride, 21–22, 59, 197
Prism, 207–208
Propane, 93, 286
Propeller, 191, 193, 196
Protozoa, 245
Providence of God, 16–17, 162, 253
Pulleys, 178

Q

Quantum Microscope, 103
Quark, 104–105, 106–107
Quartz, 103, 168
Quasar, 39–40

R

Radiation, 152, 154–156, 261
Radiators, 156
Radio Waves, 206
Rain, 17, 44, 65, 117, 124, 154, 238, 242, 248, 260, 267–271, 284, 288
Rainbow, 207–209
Rarefaction, 218
Raw Material, 13
Reaction, 116, 118, 124, 128, 144, 190–191, 193, 196, 240
    Chemical Reaction, 116, 118, 128, 144, 240
Recoil, 190
Reflection, 35, 155, 204, 208–214, 243
Refraction, 208
Repentance, 69, 71, 285–287
Resistor, 165–166

Richter Scale, 69
Ring of Fire, 68–69, 72, 77
River, 12, 17, 64, 82, 85, 124, 158, 161, 212, 230, 232, 238, 268, 280, 286
Rocket, 25, 27–29, 56, 146–147, 196–197
Rocks
    Igneous Rocks, 84
    Metamorphic Rocks, 84
    Sedimentary Rocks, 84, 91, 133
Rockwell Hardness, 129

S

Salvation, 82, 224
Sandstone, 12, 84
Saturation Point, 120
Saturn, 25, 29, 31, 41, 55–56
Seasons, 41, 44–47, 64–65, 69, 151
Seismograph, 69, 72, 77
Seismologist, 72, 77
Semi-Conductor. *See Conduction*
Silica, 84, 118, 132, 134, 136
Silicon, 25, 95, 103, 107, 108, 133, 135, 168, 215
    N-Type Silicon, 215–216
    P-Type Silicon, 215
Silicone Carbide, 135
Silk, 135
Silver, 93, 112, 128, 153, 168
Sin, 49, 64, 69, 71, 78, 121, 213, 224, 243, 249–250, 285–288
Sky, 12, 14–16, 26, 35, 38, 40, 43–45, 50, 75, 87, 89, 185, 207, 242, 259, 262, 268–271, 280, 283–284
Snow, 65, 113, 240–243, 270, 278–279
    Snowflakes, 241–242
    Snowstorms, 278–279
Sodium, 108, 112, 117
Soil, 82, 182, 242–252, 286

GOD MADE THE WORLD

Chalky Soil, 247, 249, 250
Clay Soil, 247–248, 250, 252
Fertile Soil, 246
Loamy Soil, 247, 249, 250
Mollisol Soil, 244
Peaty Soil, 247–248, 250
Sandy Soil, 182, 247, 248, 250, 252
Silty Soil, 247, 249, 250
Soil Types, 250
Solar Cells, 215
Solar Power, 96, 214–215
Solar System, 25, 28–29, 30, 40, 50, 56, 87
Solids, 112–114, 125–127, 131–132, 151–152, 196, 207, 209, 214, 217, 222–223, 231, 241, 262, 285
Solvents, 119
Sound, 143, 144, 216–219, 221–224, 274
South Pole, 66, 156, 158–159
Speakers, 221–222
Star, 14–17, 25, 27–29, 33–40, 42–44, 47, 58, 64, 75, 78, 81, 106, 173, 203, 212, 262
Static Cling, 103
Steel, 80, 117–118, 126, 128–129, 132–133, 135, 152–153, 157–158, 204, 233
    Knife Steel, 129
Storms, 17, 40–41, 259, 265, 274, 280–282, 284–288
Sun, 12, 14, 17, 19, 25, 27–31, 33–36, 39–45, 47–51, 53–57, 64, 87, 88, 103–104, 106, 148, 154–155, 184–186, 202–204, 210, 252, 261–262, 266–267, 269, 270, 275–276, 287
Superbolt, 161
Supernovas, 36
Surface Tension, 234, 239

T

Technology, 172, 197
Tectonic Plates, 72, 76, 83
Teflon, 108–109, 136
Temperature, 31–33, 65–66, 76, 113–114, 132, 149–152, 171, 215, 261, 264–265, 269, 271, 273, 275–277, 287
Tensile Strength, 131, 135
Theory, 93, 145, 147
Thermal, 143–144, 149, 153, 269–270
Thermodynamics, 149
Thrust, 193, 196
Tides, 56, 84, 96
Titan, 25
Titanium Alloy, 135
Tooth, 129
Tornadoes, 41, 265, 282–285, 287
Transparent, 208
Trees, 20, 64, 74, 79, 81, 92, 97, 98, 105, 107, 130, 212, 220, 224, 239, 243, 247, 249–250, 273, 286
Tsunami, 70, 73–74
Tungsten, 128–129, 133, 135
Typhoon, 280

U

Ultrasound, 223
Ultraviolet, 206, 261
Uranium, 128
Uranus, 29, 31, 56

V

Vegetation, 32, 74
Velocity, 181–182
Venus, 25, 29, 31–32, 55–56
Volcanoes, 68, 72–78, 80–81
    Volcanologist, 77
Voltage, 163–165, 168
Volume
    Mass, 83, 114–115, 125–126, 164, 244

INDEX

Sound, 221–223

Vredefort Crater, 89

W

Water, 12, 19, 34, 44, 52, 56, 59, 64, 69, 81, 84–85, 87, 91–93, 106, 112–114, 116–126, 128, 133–134, 149, 151–155, 160, 164–165, 168–169, 171–173, 177, 190–192, 205–209, 212, 217–219, 229–245, 247–249, 251–252, 263, 265, 267–271, 275–276, 278–280, 282–283, 286, 289

Waves, 205–207, 216–219, 221–223
- Amplitude, 205
- Electromagnetic Waves, 155–156, 202, 205–206
- Frequency, 205–206, 219, 222–223
- Resonant Frequency, 222
- Sound Waves, 216–218, 221–223
- Wave Energy, 205
- Wave Speed, 205
- Wavelengths, 155, 205–210, 214, 219

Weather, 81, 251, 259–260, 264–266, 271, 273–274, 277, 280–281, 287–288
- Weather Satellites, 287

Weight, 35, 56, 114, 126–127, 131, 133, 135, 150, 161, 202, 232–233, 262

Wheel, 148, 178, 181, 187–189, 193, 281

Whittle, Frank, 193

Wildfire. *See Fire*

Winds, 17, 20, 41, 53, 96, 144, 146–147, 160, 173, 195, 235, 242, 252, 259, 261, 266–267, 270, 277, 280–283, 287

Wisdom, 15, 17, 19–20, 38, 51, 59, 63, 82, 101, 105, 112, 115, 136, 154, 191, 201–203, 211, 219, 289

Wood, 13, 17, 98, 113, 114, 125, 126–128, 130–133, 153, 155, 160–161, 168, 170, 191, 219, 232, 239, 282, 286

Wool, 153

Worldwide Flood, 17, 64–65, 68, 76, 82–83, 84–85, 87–93, 190

Wright, Wilbur and Orville, 192

X

X-Ray, 155, 206, 261

Xenon, 116